Biobanking and Regenerative Medicine

Biobanking and Regenerative Medicine

Editor

David T. Harris

MDPI • Basel • Beijing • Wuhan • Barcelona • Belgrade • Manchester • Tokyo • Cluj • Tianjin

Editor
David T. Harris
University of Arizona
USA

Editorial Office
MDPI
St. Alban-Anlage 66
4052 Basel, Switzerland

This is a reprint of articles from the Special Issue published online in the open access journal *Journal of Clinical Medicine* (ISSN 2077-0383) (available at: https://www.mdpi.com/journal/jcm/special_issues/biobanking).

For citation purposes, cite each article independently as indicated on the article page online and as indicated below:

LastName, A.A.; LastName, B.B.; LastName, C.C. Article Title. *Journal Name* **Year**, *Article Number*, Page Range.

ISBN 978-3-03943-539-5 (Hbk)
ISBN 978-3-03943-540-1 (PDF)

Contents

About the Editor

David T. Harris is a graduate of Wake Forest University in Winston-Salem, North Carolina, where he obtained a Bachelor of Science degrees (cum laude) in Biology, Mathematics and Psychology in 1978. He earned a Masters of Medical Sciences (summa cum laude) from Bowman Gray Medical School in 1980 and his Doctorate in Microbiology and Immunology (magna cum laude) from Bowman Gray Medical School in 1982. From 1982 to 1985, Dr. Harris was a post-doctorate fellow at the Ludwig Institute for Cancer Research in Lausanne, Switzerland. In 1985, he joined the faculty at the University of North Carolina-Chapel Hill as a Research Assistant Professor in the Department of Medicine. In 1989, Dr. Harris joined the faculty at the University of Arizona in Tucson as an Associate Professor in the Department of Microbiology & Immunology. In 1996, Dr. Harris was promoted to Professor of Immunology. Dr. Harris established the first cord blood bank in the USA in 1992. He currently serves as the Executive Director of the University of Arizona Biorepository, as Quality Director of the GMP Laboratory, as a member of the Arizona Cancer Center, a member of the Children's Research Center, and as a member of the Arizona Arthritis Center. Dr. Harris's research interests include stem cells and regenerative medicine, cancer research/stem cell transplantation and gene therapy. Dr. Harris has founded five companies while at the University of Arizona, namely Cord Blood Registry, Inc.; ImmuneRegen BioSciences, Inc.; QuReGen, Inc., AdiCyte and Family Cryo Bank.

Journal of
Clinical Medicine

MDPI

Editorial

Biobanking and Regenerative Medicine: An Overview

David T. Harris

Department of Immunobiology, Arizona Health Sciences Centre, University of Arizona, Room 6122, P.O. Box 245221, Tucson, AZ 85721, USA; davidh@email.arizona.edu; Tel.: +1-520-626-5127

Received: 29 May 2018; Accepted: 29 May 2018; Published: 31 May 2018

Regenerative medicine and tissue engineering play significant roles in the treatment of currently intractable conditions such as chronic heart failure, stroke, chronic osteoarthritis, and other maladies. Regenerative medicine and tissue engineering generally depend on the utilization of stem cells to treat patients but may also utilize mature cells that would not normally be considered as stem cells (e.g., skin). Stem cells (like mature cells) may be obtained from many sources in the body including bone marrow, cord blood, cord tissue, adipose tissue, etc. Although stem cells are often used in therapy immediately upon isolation, in many circumstances the stem and progenitor cells will be harvested, processed and banked frozen until a later time. Biobanking is a convenient alternative to same-day therapeutic use, in that it allows for patient recovery (e.g., from liposuction or surgery), provides time to identify the best treatment options, and may allow for multiple interventions without additional patient inconvenience or risk.

This Special Issue will address the topic of biobanking and how it fits into regenerative medicine. Topics such as stem cell banking (e.g., cord blood, cord tissue, bone marrow, adipose tissue methodology), utilization of biobanked stem cells in pre-clinical and clinical trials, and mature cell biobanking and utilization in animal models and clinical trials (e.g., cardiomyocytes and blood vessels) will be described. Special emphasis will be put upon the role that biobanking plays in clinical therapy, precision medicine and "big data".

The establishment of biobanks has its origin in the laboratories that established repositories of tumors and other cell lines at the turn of the last century. These facilities were generally designed for local research use, although occasionally cell lines might be shared between laboratories on a limited basis. With the advent of stem cell transplantation for the treatment of blood borne cancers such as leukemia, it became commonplace to harvest and bank "back-up" bone marrow in case of treatment failure. These biobanks were once again local in nature, and generally not maintained for more than a couple of years at a time. As the specimens were patient-related, in general the samples were not shared between investigators. As the national research enterprise grew based on increasing federal dollars the value of well-established and well characterized cells, tissues and lines became more and more important. Upon this realization several public and private entities were created to fill this need, such as the American Type Culture Collection, the National Institutes of Health biospecimen service and the Coriell Institute. Access was often and still is limited, often requiring some form of financial remuneration/reimbursement. The discovery of stem cells in leftover umbilical cord and placental blood in the 1980s, and that it could be used in place of bone marrow for transplantation, led to the establishment and rapid expansion of stem cell banks worldwide. Over the course of the past 20 years more than 4 million cord blood samples alone have been biobanked in the US, and more than 40,000 samples have been thawed and used for transplant and regenerative medicine applications. Finally, the combined interest in precision medicine and big data, along with necessary clinical annotation of biospecimens, has led to an even greater demand for high quality, clinical grade biospecimens both for research and for clinical use in regenerative medicine and tissue engineering.

Although bone marrow banking is not routinely performed, cord blood banking for use in transplant and regenerative medicine has also led to the beginnings of cord tissue banking for future use in regenerative medicine and tissue engineering. Banking is done in the frozen state where samples can be stored indefinitely, as opposed to cold storage banks as done for red cells. For adults without access to their own cord blood collected at birth, adipose tissue banking, which is a rich source of mesenchymal stem cells (MSCs), has recently begun with reasonable success. In fact, frozen adipose tissue has been thawed after as long as 3 years in storage and used to successfully treat more than 200 patients.

Biobanking could be applied to almost any cell or tissue if proper methodology is employed. That is, it is technically feasible to freeze sheets of cardiomyocytes for cardiovascular applications, corneal limbal cells for ophthalmic applications, and endothelial cells for construction of vascular grafts. Biobanking can be advantageous in both the autologous and allogeneic settings, to reduce costs, to personalize therapies if needed, and to reduce patient inconvenience. In the autologous setting the collection and banking of biospecimens can inconvenience the patient only once, with multiple aliquots being set aside for future use. The biospecimen can be collected when the patient is at their youngest and healthiest, so that the cells are most optimal for use in therapy at any time in the future. In addition, it reduces the concerns about disease transmission and immune rejection. In the allogeneic setting it can permit selection of the most ideal biospecimen donor when personalized therapies are not needed. Young and healthy donors free of disease or other medical issues can be utilized, biospecimens expanded into hundreds if not thousands of therapeutic aliquots, and then placed at various banking sites around the country (or world) where they can be immediately available when needed. Creation of large autologous biospecimen banks (e.g., cord blood banks) can also permit clinical trial tailoring to specific patients with certain diseases or indications that shortens time to treatment, rapidly fills patient recruitment quotas and increases the probability of positive treatment outcomes.

There is another consideration to remember concerning biobanking and regenerative medicine; precision medicine. Large scale biospecimen banking in conjunction with highly annotated clinical data for each biospecimen is crucial to identifying optimal patient demographics, therapeutic approaches for specific patient subgroups, and laying the foundation for novel discoveries based on interrogation of the big data derived from this approach. However, to protect patient identity and confidentiality it is necessary to de-identify the biospecimens. This task can be accomplished via a clinical data warehouse (CDW) using bar codes linked to patient medical record numbers (MRNs), and MRNs linked to patient electronic medical records (EMR). The biobank itself remains blind to patient identity but is able to access patient medical records and demographics. Biospecimens if collected and stored properly may be used for both therapy and research. That is, large samples such as cord blood or adipose tissue may be later removed and used for patient treatment. However, if multiple small aliquots of the specimen are also stored those "bullets" can be used for research and interrogative purposes to determine patient qualifications for trials and improved outcomes from such trials, along with providing specimens for research interrogation that produces big data that can be the source of novel discoveries and additional therapies.

In conclusion, establishment of a biobanking enterprise can be a valuable asset for regenerative medicine. The biobank can be a source of materials for therapy and for research and development. In addition, annotation of each biospecimen (in part or in whole) with relevant patient demographics and medical data can be the source of big data that leads to better patient outcomes and discovery of new therapeutic approaches. Biobanking should not be limited solely to stem and progenitor cells, as mature and differentiated cell populations can also be medically beneficial (e.g., cardiomyocytes). The biobanking approach be constrained by whether a target population is a single cell solution, as new approaches to the cryopreservation and thawing of tissues and seeded biomaterials have proven successful and increase the utility of the biobanking facility. Investment in the biobanking endeavor can have large cost recoveries as the foundation for a successful regenerative medicine and tissue engineering program.

Conflicts of Interest: The author declares no conflict of interest.

Journal of
Clinical Medicine

MDPI

Article

Long-Term Biobanking of Intact Tissue from Lipoaspirate

Michael S. Badowski *, Angela Muise and David T. Harris

AHSC Biorepository, 1501 N. Campbell Ave., AHSC 6122, PO box 245221, The University of Arizona, Tucson, AZ 85724, USA; muisea@email.arizona.edu (A.M.); davidh@email.arizona.edu (D.T.H.)
* Correspondence: Badowski@email.arizona.edu; Tel.: +1-520-626-5125

Received: 6 January 2019; Accepted: 26 February 2019; Published: 8 March 2019

Abstract: Autologous fat grafting has now been extensively and successfully performed for more than two decades. Although most adipose grafts and adipose-derived MSC therapies are done with fresh tissue, cryopreservation of tissue allows for much greater flexibility of use. Over the course of five years, 194 cryopreserved adipose samples were thawed and then returned to the collecting physician for subsequent autologous applications. Samples were stored with a mean cryogenic storage time of 9.5 months, with some samples being stored as long as 44 months. The volumes of tissue stored varied from 12 cc to as large as 960 cc. Upon thawing, the volume of recovered whole adipose tissue averaged 67% of the original amount stored for all samples, while the samples that were stored for longer than one year averaged 71%. Recovery was not found to be a function of length of time in cryopreservation. No significant relationship was found between tissue recovery and patient age. While an average recovery of 67% of volume frozen indicates that the use of banked and thawed tissue requires a larger amount of sample to be taken from the patient initially, an experienced clinician easily accomplishes this requirement. As cryopreservation of adipose tissue becomes more commonplace, physicians will find it helpful to know the amount and quality of tissue that will be available after thawing procedures.

Keywords: adipose tissue; cryopreservation; autologous; fat grafting; cryogenic storage; lipoaspirate; cosmetic surgery; regenerative medicine

1. Introduction

Uses for autologous fat from lipoaspirate have greatly increased over the years. Autologous fat grafting is now commonplace for reconstructive, cosmetic, and regenerative medicine applications [1–3]. A variety of cellular and other therapies can include use of whole lipoaspirate, enzymatically derived stromal vascular fraction (SVF), mechanically derived microfat, or cultured and expanded adipose derived MSCs. While these and other uses become more and more standard, they still suffer some inefficiencies due to the need for fresh adipose harvests. Patients that require more than one treatment over timeframes longer than one day often need a separate harvest of lipoaspirate material for these multi-faceted interventions. One method to alleviate this issue is the simple cryopreservation of adipose tissue. The storage of adipose tissue via cryopreservation allows for the physician to dip into a store of therapeutically useful material at will. Without the need for additional adipose harvests the patient enjoys reduced morbidity, discomfort, and cost, while the treating physician has more flexibility over the course of treatments and potentially greater overall efficiency of clinical operations.

Adipose tissue has been used extensively and successfully for decades. First described late in the 19th century by Neuber [4], it was only minimally used throughout most of the 20th century. This lack of widespread adoption may have been due to poor performance, as "blocks" of adipose tissue were usually transferred with little regard for the required vascularization [5]. Only with the advent of liposuction in the 1980s [6,7], and Coleman's work in the 1990s [8], did our current concept of adipose tissue transfer emerge. The increases that were seen after Coleman's systemization of the techniques in 1995 are such that the grafting technique is now commonplace. In the intervening twenty-plus years, a greater understanding of adipose grafting has come from improved harvest techniques, increased knowledge of tissue handling, and a more experienced group of clinicians. In addition to clinical use of freshly harvested tissue, better cryopreservation protocols [9–11] and greater experience with post-thaw uses have made an equally strong case for cryopreserved adipose graft utility. Although most adipose grafts are still done with fresh tissue, the cryopreservation of tissue allows for much greater flexibility of use. A large variety of applications are now common in cosmetic uses [12,13] and stem cell therapies [14–16]. Many additional experimental applications show great promise—from cardiac repair [17,18] to endocrine therapies [19,20]. More applications are likely forthcoming as the cryopreservation of adipose tissue becomes more commonplace. As these new applications are being developed and refined, it will be important for each of these new therapies to ensure that there is enough raw lipoaspirate to use. Whatever the therapy, the amount of available lipoaspirate will affect the number of cells that are derived from the tissue, and necessarily the potential cell dosage available to the eventual patient. Therefore, the scope of this paper largely deals with raw lipoaspirate and its availability after cryopreservation, as it will be a factor in these therapies.

One advantage of adipose tissue is that the physician can use whatever amount is needed for treatment. Small-scale cosmetic applications include treatment for facial rejuvenation [12,21], fine lines and wrinkles on face and hands [22,23], and scar repair [24,25]. Larger volumes are needed for applications, such as breast reconstruction or enhancement [25,26] or other large area sculpting/remodeling. The amount of tissue needed for different applications can be exceedingly variable. While several hundred milliliters might be used for larger defects or for reconstruction, some uses will require far less. It therefore behooves the clinician to know how much fat would be a useful amount when it is time for the tissue to be applied. Many practitioners presently harvest adipose tissue and use it for reinjection during the same office visit [27]. However, it has been clear for some time that the cryopreservation of adipose tissue for future use is not only a viable option, in many ways it is advantageous. The adipose tissue that was collected from a single harvest can be cryopreserved into multiple aliquots of varied sizes. This approach reduces not only cost to the patient, but also patient discomfort, morbidity, and possible complications. As a general principle, cells that are younger are more healthy, hardy, and generally more useful. As patients age, it would be beneficial for physicians to have a bank of tissue from which they could draw. If these frozen tissues are adipose tissue aliquots from the same patient (only younger), so much the better [28]. As adipose tissue is a rich source of mesenchymal stromal cells (MSCs) and other cells future applications in cellular and regenerative medicine are amendable. As such, this work is meant to act a precursor to more advanced regenerative medicine research. Since the full utilization of freshly harvested tissue is not always possible, the amount that is available at a later date is likely to be an important factor in planning future therapies and treatments.

Cryopreservation of a large amount of tissue allows the medical professional to sample a large store of tissue whenever additional adipose tissue is needed [11]. This approach obviates the need for additional patient harvests with their associated cost and trauma. This may be extremely useful for those patients that may have less than optimal healing. Diabetics, the elderly, rheumatoid arthritics, or others with chronic wounds and ulcerated conditions may benefit the most from this approach as multiple treatments may be indicated. To aid the clinician in determining the amount of adipose tissue that should be harvested for cryopreservation, this study analyzes the recovery yields supporting the utility of cryopreserved adipose tissue for future clinical use.

2. Materials and Methods

2.1. Tissue Collection and Processing

Lipoaspirate samples were collected from male and female adult patients undergoing voluntary liposuction. All of the samples were obtained with written consent from the donors according to the instructions from the local institution review board at the University of Arizona. No paid compensation was given to subjects. Tissue was collected into sterile vessels and then allowed to separate by sedimentation. In this manner blood, tumescent fluid and oils were removed from tissue. The separated tissue was packaged into syringes and then shipped via overnight or local courier. Upon receipt at the processing laboratory, syringes were placed upright at room temperature for a further round of sedimentation. Any remaining fluids were removed from each syringe along with a small amount of adipose tissue into a sterile tube. The samples were pooled into a single sterile tube when multiple syringes arrived for processing. This extract was used for viability and sterility testing. If no tumescent fluid were present in the packaged syringes, a small amount of adipose tissue was removed and then vigorously mixed with sterile PBS to facilitate testing.

2.2. Sterility Testing

Sample liquids were streaked with a 10 μL sterile loop onto growth plates. Using a sterile loop, the sample was streaked onto MacConkey agar, Sabaroud-dextrose agar, and 5% sheep's blood tryptone soy agar plates (University of Arizona Bio5 Media Facility, Tucson, AZ, USA). All of the agar plates were incubated and monitored for seven days in a 37 °C dry incubator. The appearance of colonies at any point during the seven days was scored as microbial growth present. In lieu of testing on plates, some of the samples were tested with the BacT/ALERT system (Biomerieux, Marcy l'Etoile, France). For these assays, sample liquid was loaded into a sterile syringe with needle and injected into the culture liquid through a rubber septum. Both the aerobic and anaerobic media bottles were kept for seven days at 37 °C. Any growth within seven days was scored as positive. The sterility results were reported to the physician but are not part of this study.

2.3. Histology

Small pieces of adipose tissue were placed on a microscope slide. A second slide was used to smear the tissue across the slide. It was then fixed in methanol for 30 s and placed in 0.1% Crystal Violet for at least one hour. When stain had permeated completely though tissue, the slide was rinsed gently by dipping in clean water until all excess stain was removed. Tissue preparation was allowed to air dry. A cover slip was placed over stained tissue, secured in place with Histomount, and then allowed to dry for a least a day. The slides were viewed on a standard brightfield light microscope.

2.4. Cryostorage

Adipose tissue was treated, as previously described [11]. Tissue was loaded into sterile cryostorage bags (Origen Biomedical, Austin, TX, USA) and cooled to 4 °C. Ice cold cryoprotectant was added to achieve a final concentration of 5% DMSO, 1% dextran-40 (Protide Pharmaceuticals, Lake Zurich, IL, USA), and 1% human serum albumen (Octapharma, Lachen, Switzerland). After mixing, the bags were sealed into secondary sterile overwrap. Bags were frozen according to a preprogrammed cooling curve in a controlled rate freezer (Custom Biogenic Systems, Romeo, MI, USA) and then stored in vapor phase LN2 dewar at −180 °C.

2.5. Thawing of Frozen Adipose Tissue

Cryostorage bags were rapidly thawed in a 37 °C water bath and washed twice with cold lactated Ringer's solution. Wash liquid was separated from tissue and the volume of tissue measured in the same manner as samples prior to cryopreservation. Washed tissue was then transferred to 20 cc or 60 cc packaging syringes. Tissue was loaded in a 1:1 ratio, with transport buffer consisting of lactated Ringer's solution with 1% human serum albumen. Tissue was kept refrigerated in a temperature-controlled shipping container overnight or local courier delivery to the administering clinician.

3. Results

Adipose Tissue Recovery after Prolonged Cryopreservation and Thawing

Samples, or aliquots of samples, were in cryostorage for an average approximately 9 $\frac{1}{2}$ months. A total of 181 samples were thawed after various lengths of time following cryopreservation. The samples had been stored for periods ranging from 13 days to more than 44 months. Long-term samples were considered to be in storage for more than one year. 38 samples met this long-term criterion. After the thaw procedures were complete, the amount of tissue recovered was compared to the amount of tissue originally cryopreserved in each sample or aliquot. Due to original removal of waste liquids (blood, tumescent fluid, oil) prior to cryostorage, those volumes were not included in the amount of tissue stored. The average recovery yield of cryopreserved tissue volume was 67% of the original stored amount. Tissue recovery ranged from a low of 21% to a high of 100%. The large majority of samples fell between 50% and 80%. For samples that were classified as long-term, i.e. those in cryostorage for longer than one year, the yields are slightly higher. An average of 71% of tissue volume was recovered after thaw procedures. Figure 1 shows the distribution of recovery yield for these long-term samples. There appears to be no correlation of tissue recovery with length of time in cryostorage.

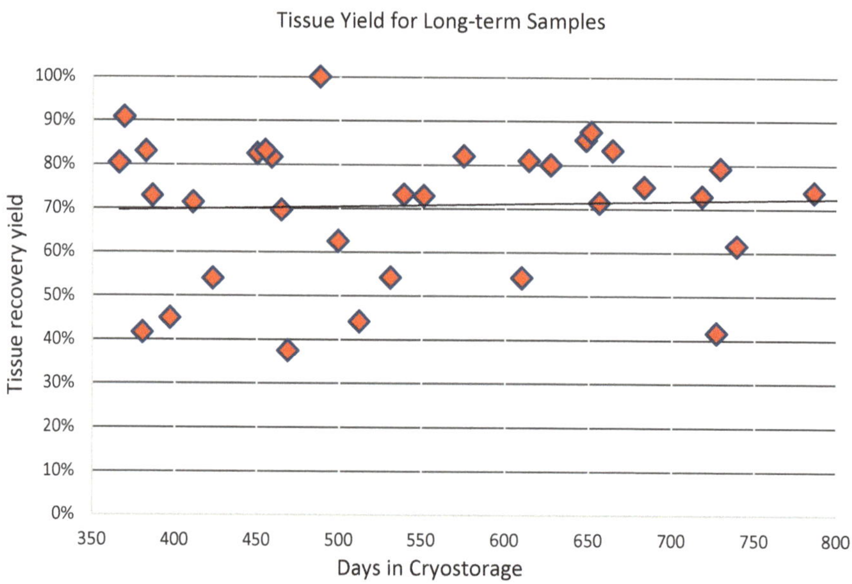

Figure 1. Long-term yield. Adipose tissue was stored for 38 patients for times longer than one year. Upon thaw an average of 71% of original tissue volume was available for subsequent use.

Although the majority of patients in this study ranged from 50 to 70 years of age, the entire patient population was from 27 to 81 years old. The yield of adipose tissue recovery after thawing appears to have very little dependence on the age of the patient. Linear regression analysis shows that, for each additional decade of age, the reduction in recovery is less than one percentage point. Figure 2 shows the distribution of post-thaw tissue yield and its relationship to donor age.

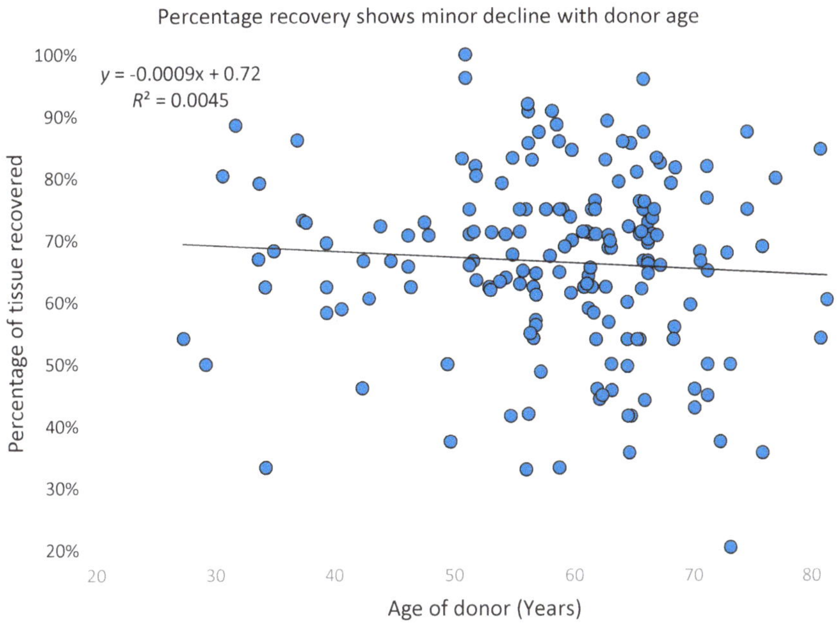

Figure 2. Percentage recovery shows little decline with age. The age of adipose tissue donors ranged from 27 years to 83 years. Regardless of donor age, there was no significant decrease in thawed tissue yield after cryopreservation.

The adipose samples that were cryopreserved in this study ranged from 12 cc to 960 cc (mean = 242 cc) of the total sample. This volume included some waste materials, such as blood, oils, and tumescent fluid, which were packaged along with the adipose tissue into the collection syringes. These waste fractions were discarded prior to adipose cryopreservation. Samples that were smaller than 12 cc were excluded, but there was no upper limit on the collected volume. Therefore, the amount of lipoaspirate cryopreserved was largely dependent upon how much the physician harvested. Since large samples were stored in multiple aliquots, the sample was, thereafter, available for multiple autologous uses. Some physicians had taken out aliquots of sample on multiple occasions for multiple post-thaw usages. Figure 3 shows the recovery percentages varied less within a set of aliquots from the same donor.

Tissue prepared onto thick sections and stained with crystal violet was viewed on a light microscope. Samples from fresh adipose tissue and cryopreserved post-thaw adipose tissue were essentially the same. While the large-scale structure of the adipose tissue is disrupted during lipo-extraction, much of the same structures are observed. Intact pieces of tissue, outlines of intact adipocytes, collagen strands and matrix, and density of nuclei can be seen. There appears to be no gross differences in slides that were prepared from fresh or post-thawed adipose tissue. Figure 4 shows two examples of each fresh tissue and post-thawed tissue preparations. It displays that the prepared tissue essentially presents the same histology characteristics from both freshly harvested and cryopreserved/thawed tissue. While the authors understand that H&E staining is the norm for basic histology, crystal violet (CV) was chosen instead of H&E staining for superior visualization of tissue under our conditions. CV has the property of strongly staining both nuclei and intercellular matrix, which is not as robustly seen in H&E. Intracellular structure is more prominently stained with H&E in most cells. However, intracellularly adipocytes are nearly completely all fat filled vacuoles. A characteristic of this feature is that it does not stain with H&E. A strong outline of the shape of the cells can be seen more easily with CV than H&E. These factors led us to choose CV over H&E staining.

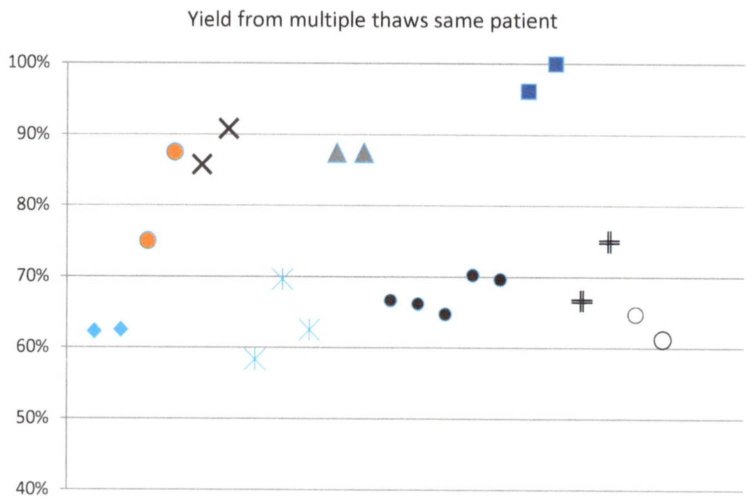

Figure 3. Multiple thaws from same patient are similar. Large adipose samples were stored in multiple smaller aliquots. Individual patients ($n = 9$) are shown here as markers of the same style. Tissue yield from thaws procedures on different days show similar recovery levels.

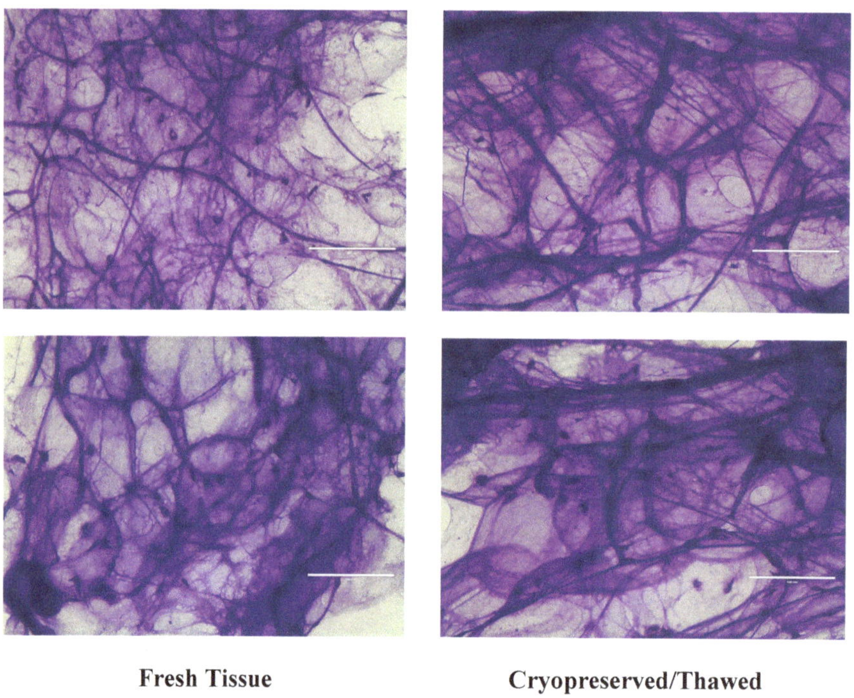

Figure 4. No gross differences between fresh and frozen histology. Adipose tissue was mounted on slides and then stained with crystal violet. No differences in structure are observed in fresh or cryopreserved/thawed tissue.

4. Discussion/Conclusions

Adipose tissue has been shown to have useful for a great many conditions and it has promise for a great many more. It is easy to harvest, easy to handle, and, for many patients, rarely is it in shortage. It provides a rich source of adipose derived MSCs, pericytes, endothelial cells, pre-adipocytes, and others, which are known to have large numbers of current and potential therapies. All else being equal, younger cells are better than older cells [28–30]. Therefore, it would seem beneficial for patients to have a bank of stored autologous adipose tissue taken while the patient is young. From this cryopreserved collection, the aliquots can be taken out periodically for the cosmetic needs, regenerative medicine, cellular therapies, and reconstructive applications, which the patient may require in the future.

In this study, we have seen that the average yield of tissue recovery after thawing is 67% for all samples (21% to 100%). For long-term samples we observed that the average tissue yield recovery is 71%. While slightly higher than the 67% that was found in all samples, it was not found to be significantly higher. The practical application is that, after the thawed tissue is washed and recovered, it allows the physician to easily predict the amount of adipose tissue available for use. A conservative approach allows for a physician to estimate that one-half to two-thirds of the original banked amount will be available for fat transfer applications. Physicians can make this useful estimate for all additional tissue frozen, regardless of aliquot size, overall harvest volume, or length of time in storage. Cryopreservation of cells has been commonplace for decades. Indeed, the samples of adipose tissue act similarly to other properly cryopreserved cells. Essentially, they can be stored indefinitely. The samples remained cooled at liquid nitrogen temperatures of approximately -185 °C. At this temperature essentially, all biological activity is stopped. Without this complete cessation we may expect to see degradation over time. However, properly stored under appropriate conditions, no reduction of recovery yield is seen as the time in cryostorage increases – even over a time frame of years.

An average recovery of 71% for long-term samples of frozen adipose tissue indicates that the use of banked and thawed tissue initially requires a larger amount of sample be taken from the patient. Any clinician that is competent in harvesting technique easily accomplishes this requirement. Our samples in this study were stored in cryopreservation bags that allowed storage of aliquots of adipose tissue of 24 cc or 56 cc per bag. Some physicians had stated that a portion of the fresh lipoaspirate had been used that same day and the remainder had been sent to the processing lab for cryopreservation. However, data regarding the number of these cases and the amount of tissue for each purpose is not available. Outside of this study, we have stored adipose tissue in other types and sizes of vessels that resulted in aliquots of tissue, ranging from 2 cc to more than 200 cc. A wide variety of cryostorage bag sizes are available from our supplier (Origen Biomedical, Austin, TX, USA), as well as other suppliers. This allows for the adipose tissue to essentially be stored in any volume. It seems reasonable that large volumes of tissue are best cryopreserved in multiple aliquots. Multiple aliquots allow the clinician to just thaw the volume that is needed for the present application, while holding the remainder in reserve. While it is possible to refreeze adipose tissue and still utilize it upon a second (or third) thaw, there is an observed reduction in both cell viability and overall volume yield [31]. Once a clinician has an approximate idea of the amount of tissue that is required for a given procedure, the amount of tissue aliquots to be thawed can be predicted. In this way, the clinician is assured of not running out of tissue or of thawing too many bags that will go to waste.

As the adipose tissue is processed and cryopreserved, a small amount of tissue is lost through the normal manipulations that are required. For example, tissue that cannot be easily removed from a syringe or an amount of tissue may become stuck in tubing is unrecoverable. The amount that is lost in the unloading of the collection syringes and the transfer to cryostorage bags is minimal. However, some adipocytes burst and are destroyed during the various processes involved. Moving fragments of lipoaspirate through a small orifice, exposing cells to pressure differences and sheer stresses, changes in osmotic pressure with the cells during cryopreservation and thaw, washing, and centrifugation can all have an effect on the tissue. Taken together, these processes affect a change in the overall volume of the recovered sample. It was noticed that, most often, amounts of tissue were unrecoverable during the thaw and wash procedures and the subsequent transfer of tissue to packaging syringes for transport. Ensuring that the size of cryostorage bag was not overly large as compared to the volume of tissue stored minimized this loss. It was noticed early on that lowest yields were often from cryostorage bags that were filled to less than capacity with adipose tissue. Tissue can be stored in any amount and cryostorage bags come in many sizes. However, more tissue was unrecoverable or lost when the bag was too large for the tissue stored. There was no significant correlation between the total amount of tissue stored and the percentage yield of individual samples.

There are different reasons for storage of adipose tissue; e.g., for the near-term use for cosmetic purposes. Many of our current tissue retrievals have been for reinjection into the face or hands. Other observed uses have been for larger scale remodeling/reconstruction, such as breast or buttocks. Another reason for the storage of tissue is that adipose tissue is abundant with MSCs, pericytes, and other cells. This potent tissue will, no doubt, be even more useful in the future as yet unknown, regenerative medicine applications are discovered. It is also known that, the younger the stem cells are, the more potent and more useful they will be [16,28–30]. Therefore, it behooves patients to collect a sample of adipose tissue and store it at a time as early as is feasible and to keep it stored instead of for short-term use for cosmetic applications.

Some cellular therapies will depend upon the extraction and expansion of MSCs. For this, small amounts of tissue can be expanded to great use, provided that the viability and cellularity are acceptable. Our previous work has shown this method of cryopreservation and thaw yields viable cell counts that are appropriate for expansion and therefore potential therapies that depend on it. [11] However, in cosmetic and reconstructive cases, the amount and volume of tissue that a physician has to work with is of more importance. As the tissue was processed for thaw (washings, transfers, etc.), only the actual amount of tissue that ended up in the physician's hands was measured as the final product. As the cryopreservation of adipose tissue becomes more commonplace, physicians will find it helpful to know what will be the likely amount of tissue that will be available after the thaw procedures. The final product ready to use was sufficiently washed and the tissue yield was measured. As the tissue is packaged for use into syringes with osmotically balanced buffer, the physician can immediately use the tissue upon arrival by merely decanting the buffer.

This process is technically very simple for the physician. However, some may not choose such procedures as cryopreservation and fat graft for the concern that it runs afoul of rules and guidance from the U.S. Food and Drug Administration (FDA). It merits some attention that the FDA has produced guidance regarding "processing" of Human Cell, Tissues, and Cellular and Tissue-based Products (HCT/Ps). However, what is herein described falls under CFR Title 21, part 1271.15 (b), which states that "You are not required to comply with the requirements of this part if you are an establishment that removes HCT/P's from an individual and implants such HCT/P's into the same individual during the same surgical procedure". This exception was further clarified later in FDA Guidance in November 2017. The example of autologous adipose tissue was specifically given and the FDA states "the adipose tissue retains its original form as a connective tissue composed of clusters of adipocytes and other cells surrounded by a reticular fiber network and interspersed small blood vessels. It is then re-implanted into the same patient from whom it was removed in order to achieve the intended effect. We generally would consider the establishment removing and implanting this HCT/P from adipose tissue to qualify for the exception under 21 CFR 1271.15(b)" [32].

The amount of adipose tissue that can be harvested from any given patient is dependent upon the amount of fat on the patient. However, more adipose tissue can nearly always be taken than what usually is taken. Part of the reason for this is that many physicians are currently unaware of the value of harvesting adipose tissue for cryopreservation. Coleman's systemization of liposuction technique [8] for the medical community increased its utility to reach this current level of use of fat grafting. It is reasonable that many physicians are not yet aware of, or comfortable enough with, cryopreservation to fully utilize the practice. Therefore, cryopreservation is, as of yet, not something that most harvesting doctors consider. As more information on its value and usefulness comes out, cryopreservation will undoubtedly become more popular.

Author Contributions: The following authors contributed in the ways noted: conceptualization, M.S.B. and D.T.H.; methodology, M.S.B. and A.M.; software, M.S.B.; validation, M.S.B. and D.T.H.; formal analysis, M.S.B.; investigation, M.S.B. and A.M.; resources, M.S.B.; data curation, M.S.B.; writing—original draft preparation, M.S.B.; writing—review and editing, D.T.H.; visualization, M.S.B.; supervision, D.T.H.; project administration, D.T.H.; funding acquisition, M.S.B. and D.T.H.

Acknowledgments: Portions of this work were supported by AdiCyte Inc. and Celebration Stem Cell Centre.

Conflicts of Interest: The authors declare no conflict of interest.

Abbreviations

MSC	mesenchymal stromal cells
DMSO	dimethyl sulfoxide
PBS	phosphate buffered saline
SVF	stromal vascular fraction
CV	crystal violet

References

1. Zuk, P.A.; Zhu, M.; Ashjian, P.; De Ugarte, D.A.; Huang, J.I.; Mizuno, H.; Alfonso, Z.C.; Fraser, J.K.; Benhaim, P.; Hedrick, M.H. Human Adipose Tissue Is a Source of Multipotent Stem Cells. *MBoC* **2002**, *13*, 4279–4295. [CrossRef] [PubMed]
2. De Ugarte, D.A.; Ashjian, P.H.; Elbarbary, A.; Hedrick, M.H. Future of Fat as Raw Material for Tissue Regeneration. *Ann. Plast. Surg.* **2003**, *50*, 215–219. [CrossRef] [PubMed]
3. Strem, B.M.; Hedrick, M.H. The growing importance of fat in regenerative medicine. *Trends Biotechnol.* **2005**, *23*, 64–66. [CrossRef] [PubMed]
4. Neuber, G. Uber die Wiederanheilung vollstandig vom Korper getrennter, die ganze Fettschicht enthaltender Hautstucke. *Zbl F Chirurgie* **1893**, *30*, 16–17.
5. Peer, L.A. *Transplantation of Tissue*; Williams & Wilkins: Baltimore, MD, USA, 1955.
6. Illouz, Y.G. The fat cell "graft": A new technique to fill depressions. *Plast. Reconstr. Surg.* **1986**, *78*, 122–123. [CrossRef] [PubMed]
7. Fornier, P.F. Microlipoextraction et microlipoinjection. *Rev. Cir. Esthet. Langue.* **1985**, *10*, 36–40.
8. Coleman, S.R. Long-Term survival of fat transplants: Controlled demonstrations. *Aesthet. Plast. Surg.* **1995**, *19*, 421–425. [CrossRef]
9. Pu, L.L.Q.; Cui, X.; Fink, B.F.; Cibull, M.L.; Gao, D. Long-term preservation of adipose aspirates after conventional lipoplasty. *Aesthet. Surg. J.* **2004**, *24*, 536–541. [CrossRef]
10. Shu, Z.; Gao, D.; Pu, L.L. Update on Cryopreservation of Adipose Tissue and Adipose-derived Stem Cells. *Clin. Plast. Surg.* **2015**, *42*, 209–218. [CrossRef]
11. Choudhery, M.S.; Badowski, M.; Muise, A.; Pierce, J.; Harris, D.T. Cryopreservation of whole adipose tissue for future use in regenerative medicine. *J. Surg. Res.* **2014**, *187*, 24–35. [CrossRef]
12. Marten, T.J.; Elyassnia, D. Fat Grafting in Facial Rejuvenation. *Clin. Plast. Surg.* **2015**, *42*, 219–252. [CrossRef] [PubMed]
13. Bircoll, M. Cosmetic breast augmentation using autologous fat and liposuction techniques. *Plast. Reconstr. Surg.* **1987**, *79*, 267–271. [CrossRef] [PubMed]
14. Wakabayashi, K.; Hamada, C.; Kanda, R.; Nakano, T.; Io, H.; Horikoshi, S.; Tomino, Y. Adipose-derived mesenchymal stem cells transplantation facilitate experimental peritoneal fibrosis repair by suppressing epithelial–mesenchymal transition. *J. Nephrol.* **2014**, *27*, 507–514. [CrossRef] [PubMed]
15. Bai, X.; Alt, E. Myocardial regeneration potential of adipose tissue-derived stem cells. *Biochem. Biophys. Res. Commun.* **2010**, *401*, 321–326. [CrossRef] [PubMed]

16. Yang, Y.-C.; Liu, B.-S.; Shen, C.-C.; Lin, C.-H.; Chiao, M.-T.; Cheng, H.-C. Transplantation of Adipose Tissue-Derived Stem Cells for Treatment of Focal Cerebral Ischemia. *Curr. Neurovasc. Res.* **2011**, *8*, 1–13. [CrossRef] [PubMed]
17. Ohnishi, S.; Ohgushi, H.; Kitamura, S.; Nagaya, N. Mesenchymal Stem Cells for the Treatment of Heart Failure. *Int. J. Hematol.* **2007**, *86*, 17–21. [CrossRef] [PubMed]
18. Ii, M.; Horii, M.; Yokoyama, A.; Shoji, T.; Mifune, Y.; Kawamoto, A.; Asahi, M.; Asahara, T. Synergistic effect of adipose-derived stem cell therapy and bone marrow progenitor recruitment in ischemic heart. *Lab. Investig.* **2011**, *91*, 539–552. [CrossRef] [PubMed]
19. Timper, K.; Seboek, D.; Eberhardt, M.; Linscheid, P.; Christ-Crain, M.; Keller, U.; Müller, B.; Zulewski, H. Human adipose tissue-derived mesenchymal stem cells differentiate into insulin, somatostatin, and glucagon expressing cells. *Biochem. Biophys. Res. Commun.* **2006**, *341*, 1135–1140. [CrossRef]
20. Dave, S.D.; Vanikar, A.V.; Trivedi, H.L. Extrinsic Factors Promoting In Vitro Differentiation of Insulin-Secreting Cells from Human Adipose Tissue-Derived Mesenchymal Stem Cells. *Appl. Biochem. Biotechnol.* **2013**, *170*, 962–971. [CrossRef]
21. McKinney, P.; Pandya, S. Use of pubic fat as a graft for eyelid defects. *Aesthet. Plast. Surg.* **1994**, *18*, 383–385. [CrossRef]
22. Li, T.-S.; Choong, M.-Y.; Wu, C.-I.; Chen, H.-C.; Chung, K.-C. Autologous Fat Graft to Restore Nail Bed Contour Following Resection of a Subungual Glomus. *J. Hand Surg. Am.* **2011**, *36*, 726–728. [CrossRef] [PubMed]
23. Perello, R.; Agostini, T. Lipomodeling: An Innovative Approach to Global Volumetric Rejuvenation of the Hand. *Aesthet. Surg. J.* **2015**, *35*, 708–714.
24. Klinger, M.; Caviggioli, F.; Klinger, F.M.; Giannasi, S.; Bandi, V.; Banzatti, B.; Forcellini, D.; Maione, L.; Catania, B.; Vinci, V. Autologous Fat Graft in Scar Treatment. *J. Craniofac. Surg.* **2013**, *24*, 1610–1615. [CrossRef] [PubMed]
25. Kiviluoto, O. Use of free fat transplants to prevent epidural scar formation. An experimental study. *Acta Orthop. Scand.* **1976**, *47*, 3–75. [CrossRef]
26. Shakhov, A.A. Fat Transplantation and Breast Augmentation. *Aesthet. Plast. Surg.* **2002**, *112*, 323–325. [CrossRef]
27. Carraway, J.H.; Mellow, C.G. Syringe Aspiration and Fat Concentration: A Simple Technique for Autologous Fat Injection. *Ann. Plast. Surg.* **1990**, *24*, 293–297. [CrossRef] [PubMed]
28. Choudhery, M.S.; Badowski, M.; Muise, A.; Pierce, J.; Harris, D.T. Donor age negatively impacts adipose tissue-derived mesenchymal stem cell expansion and differentiation. *J. Trans. Med.* **2014**, *12*. [CrossRef] [PubMed]
29. Yang, Y.H.K. Aging of mesnchymal stem cells: Implication in regenerative medicine. *Regen. Therap.* **2018**, *9*, 120–122.
30. Alt, E.U.; Senst, C.; Murthy, S.N.; Slakey, D.P.; Dupin, C.L.; Chaffin, A.E.; Kadowitz, P.J.; Izadpanah, R. Aging alters tissue resident mesenchymal stem cell properties. *Stem Cell Res.* **2012**, *8*, 215–225. [CrossRef] [PubMed]
31. Harris, D.T.; The University of Arizona, Tucson, AZ, USA; Badowski, M.S.; The University of Arizona, Tucson, AZ, USA. Personal Communication, 2002.
32. U.S. Department of Health and Human Services; Food and Drug Administration. Same Surgical Procedure Exception under 21 CFR 1271.15(b): Questions and Answers Regarding Scope of the Exception, Guidance for Industry. Available online: https://www.fda.gov/downloads/biologicsbloodvaccines/guidancecomplianceregulatoryinformation/guidances/tissue/ucm419926.pdf (accessed on 1 January 2019).

Journal of
Clinical Medicine

MDPI

Article

Biobanking: Objectives, Requirements, and Future Challenges—Experiences from the Munich Vascular Biobank

Jaroslav Pelisek [1,*], Renate Hegenloh [2], Sabine Bauer [2], Susanne Metschl [2], Jessica Pauli [2], Nadiya Glukha [2], Albert Busch [2], Benedikt Reutersberg [2], Michael Kallmayer [2], Matthias Trenner [2], Heiko Wendorff [2], Pavlos Tsantilas [2], Sofie Schmid [2], Christoph Knappich [2], Christoph Schaeffer [2], Thomas Stadlbauer [2], Gabor Biro [2], Uta Wertern [2], Franz Meisner [2], Kerstin Stoklasa [2], Anna-Leonie Menges [2], Oksana Radu [2], Sabine Dallmann-Sieber [2], Angelos Karlas [2], Eva Knipfer [2], Christian Reeps [3], Alexander Zimmermann [2], Lars Maegdefessel [1] and Hans-Henning Eckstein [1]

[1] DZHK (German Centre for Cardiovascular Research), Munich Heart Alliance, 80636 Munich, Germany; Lars.Maegdefessel@tum.de (L.M.); HHEckstein@web.de (H.-H.E.)

[2] Department of Vascular and Endovascular Surgery, Technische Universität München, 81675 Munich, Germany; renate.hegenloh@tum.de (R.H.); s.bauer@tum.de (S.B.); susanne.metschl@tum.de (S.M.); jessica.pauli@tum.de (J.P.); nadiya.glukha@tum.de (N.G.); Albert.Busch@mri.tum.de (A.B.); Benedikt.Reutersberg@mri.tum.de (B.R.); Michael.Kallmayer@mri.tum.de (M.K.); Matthias.Trenner@mri.tum.de (M.T.); Heiko.Wendorff@mri.tum.de (H.W.); Pavlos.Tsantilas@mri.tum.de (P.T.); sofie.schmid@mri.tum.de (S.S.) Christoph.Knappich@mri.tum.de (C.K.); Christoph.Schaeffer@mri.tum.de (C.S.) Thomas.Stadlbauer@mri.tum.de (T.S.); Gabor.Biro@mri.tum.de (G.B.); uta.werthern@mri.tum.de (U.W.); Franz.Meisner@mri.tum.de (F.M.); Kerstin.Stoklasa@mri.tum.de (K.S.); anna-leonie.menges@mri.tum.de (A.-L.M.); Oksana.Radu@mri.tum.de (O.R.); sabine.dallmann-sieber@mri.tum.de (S.D.-S.); Angelos.Karlas@mri.tum.de (A.K.); Eva.Knipfer@mri.tum.de (E.K.); Alexander.Zimmermann@mri.tum.de (A.Z.)

[3] University Centre for Vascular Medicine and Department of Vascular Surgery, University Hospital Carl Gustav Carus, Dresden University of Technology, 01307 Dresden, Germany; Christian.Reeps@uniklinikum-dresden.de

* Correspondence: j.pelisek@tum.de; Tel.: +49-89-4140-5168

Received: 9 January 2019; Accepted: 12 February 2019; Published: 16 February 2019

Abstract: Collecting biological tissue samples in a biobank grants a unique opportunity to validate diagnostic and therapeutic strategies for translational and clinical research. In the present work, we provide our long-standing experience in establishing and maintaining a biobank of vascular tissue samples, including the evaluation of tissue quality, especially in formalin-fixed paraffin-embedded specimens (FFPE). Our Munich Vascular Biobank includes, thus far, vascular biomaterial from patients with high-grade carotid artery stenosis (n = 1567), peripheral arterial disease (n = 703), and abdominal aortic aneurysm (n = 481) from our Department of Vascular and Endovascular Surgery (January 2004–December 2018). Vascular tissue samples are continuously processed and characterized to assess tissue morphology, histological quality, cellular composition, inflammation, calcification, neovascularization, and the content of elastin and collagen fibers. Atherosclerotic plaques are further classified in accordance with the American Heart Association (AHA), and plaque stability is determined. In order to assess the quality of RNA from FFPE tissue samples over time (2009–2018), RNA integrity number (RIN) and the extent of RNA fragmentation were evaluated. Expression analysis was performed with two housekeeping genes—glyceraldehyde 3-phosphate dehydrogenase (*GAPDH*) and beta-actin (*ACTB*)—using TaqMan-based quantitative reverse-transcription polymerase chain reaction (qRT)-PCR. FFPE biospecimens demonstrated unaltered RNA stability over time for up to 10 years. Furthermore, we provide a protocol for processing tissue samples in our Munich

Vascular Biobank. In this work, we demonstrate that biobanking is an important tool not only for scientific research but also for clinical usage and personalized medicine.

Keywords: Munich Vascular Biobank; atherosclerosis; human vascular tissue; RIN; RNA fragmentation

---

1. Introduction

In general, biobanking is considered as a tool to collect and store biological tissue samples, mainly for research purposes. Management of a successful biobank requires a high level of collaboration and networking between various medical specialists, surgeons, biologists, and technicians to process the human specimens properly in time and to acquire not only biological but also clinical data. Biobanks are built to improve our understanding of various diseases and to help develop novel therapeutic strategies [1]. They are often the only sources of biological material available to explore uncommon human disorders, to solve the problems of heterogeneity of disease-related specific biomarkers, to determine the appropriate cut-off values, or to validate results achieved in animal experiments. Furthermore, new scientific approaches, such as next-generation sequencing or omics analyses, require high quality biospecimens and thus pose additional challenges for researchers [2–4].

In the last decades, new biobanks have been built focusing on selective human tissues, such as various cancer types or atherosclerotic specimens [1–3]. In this context, the collection of human atherosclerotic biomaterial following, for example, surgical intervention of stenotic carotid arteries or excision of aortic wall from patients with abdominal aortic aneurysm (AAA), has become a particularly important issue to improve our understanding of the underlying mechanisms leading to the formation and progression of atherosclerotic plaques and aneurysm toward vulnerable and rupture-prone phenotype. The first steps to collect vascular biospecimens from autopsies were performed in the 18th century by Rudolph Virchow, who explored inflammatory changes in atherosclerotic lesions [5]. It was not until a century later that another researcher, Russell Ross, summarized his histological observations on atherosclerotic tissue and postulated the famous response to injury theory of atherosclerosis [6]. Since then, other scientists have defined different stages of thrombus formation—occlusive thrombus, intraplaque thrombus, and transitional thrombus—and demonstrated plaque vulnerability to be a function of the reduced number of smooth muscle cells, increased inflammation, and size of the lipid core under a thin fibrous cap [7–10]. Renu Virmani et al. investigated coronary arteries from patients with sudden cardiac death and characterized additional plaque features, such as erosion, calcified nodules, fibrous cap atheroma, and fibrocalcific plaques [11]. All these studies were based on observations of large atherosclerotic tissue cohorts collected over years in the corresponding biobanks [12]. Thus, the collection of vascular tissue has clearly proven its importance and usefulness in discovering new features of atherosclerosis and factors contributing to plaque vulnerability [13–15].

The main challenge to run a biobank for accurate analyses, including next-generation sequencing and omics technologies, is to guarantee sufficient quality of the biospecimens. Universal methods for collecting, storing, and processing the obtained tissue samples are still lacking. These circumstances complicate translational insights as different biobanks are rarely directly comparable, and most studies focus on the results of their individual tissue and data collections. Thus, the number of specimen and patient data, especially long-term medical history, is always limited. In this work, we provide our long-standing experience in running a biobank of vascular tissue samples, including our analyses of tissue quality from FFPE samples, for successful and reliable expression analyses.

2. Experimental Section

2.1. Ethics Approval and Study Population

The permission to collect human carotid atherosclerotic, AAA, and peripheral aortic disease (PAD) biospecimens in our Munich Vascular Biobank was approved by the local Hospital Ethics Committee (2799/10, Ethikkommission der Fakultät für Medizin der Technischen Universität München, Munich, Germany). Written informed consent was obtained from all patients. Experiments were performed in accordance with the principles of the Declaration of Helsinki. The study population includes patients with high-grade carotid artery stenosis, PAD, and AAA, who underwent surgical intervention in our Department of Vascular and Endovascular Surgery in accordance with the current recommended guidelines [16–18].

2.2. Tissue Processing

A majority of the tissue samples were processed within two to four hours following surgical excision. According to our previous experiments and that of other researchers on RNA stability [19–23], no significant degradation of RNA is observed for up to eight hours at room temperature and 12–24 h on ice. In case of noncompliance, the time between excision and tissue processing was recorded, and these samples were used only for specific analyses where, for example, RNA stability is not the main issue.

Depending on size and quality of the tissue, consecutive segments were prepared, and the selected ones were directly frozen in liquid nitrogen and stored at −80 °C. The remaining segments were fixed in formalin overnight (minimum of 24 h) and decalcified using EDTA (0.2 M ethylenediaminetetraacetic acid in 50 mM tris(hydroxymethyl)aminomethane adjusted to pH 8.0) for a couple of days depending on the extent of calcification. Afterward, the tissue segments were embedded in paraffin and stored at room temperature. From all segments, consecutive sections of 2–3 µm were prepared and stained with haematoxylin–eosin (HE) and elastin–van Gieson (EvG) to assess the tissue morphology.

The individual histological features were graduated semiquantitatively by independent assessors as follows: (-) no staining; (+) positive staining of most cells or the analyzed components of vessel wall in a majority of specimens; (++) medium overall positive staining detected in all cells/vessel wall components in all specimens; (+++) strong overall positive staining detected in all cells/vessel wall components in all specimens; (-/+) intensity of staining varied between different samples and also within the same specimen; (-/++) heterogeneous staining results from (-) to (++).

2.3. RNA Extraction from FFPE Biospecimens by Expression of Housekeeping Genes

Total RNA was isolated from FFPE tissue sections (4 × 20 µm thickness) adjacent to the slides applied for histological characterizations using the High Pure RNA Paraffin Kit in accordance with the manufacturer's instructions (Roche, Mannheim, Germany). The amount and the purity of RNA were determined using Nanodrop 2000 (Thermo Fisher Scientific, Munich, Germany).

2.4. Analysis of RNA Quality from FFPE Biospecimens by RIN and RNA Fragmentation

RNA integrity number (RIN) was determined by Agilent 2100 Bioanalyzer and the RNA 6000 Nano Kit (Agilent Technologies, Waldbronn, Germany) in accordance with the manufacturer's instructions. In order to receive additional information about the quality of RNA, the length of the RNA fragments was evaluated from the data curve of the Bioanalyzer, calculating the maximal length and the RNA fragment size at 50% reduction of the area under the curve (Figure 1).

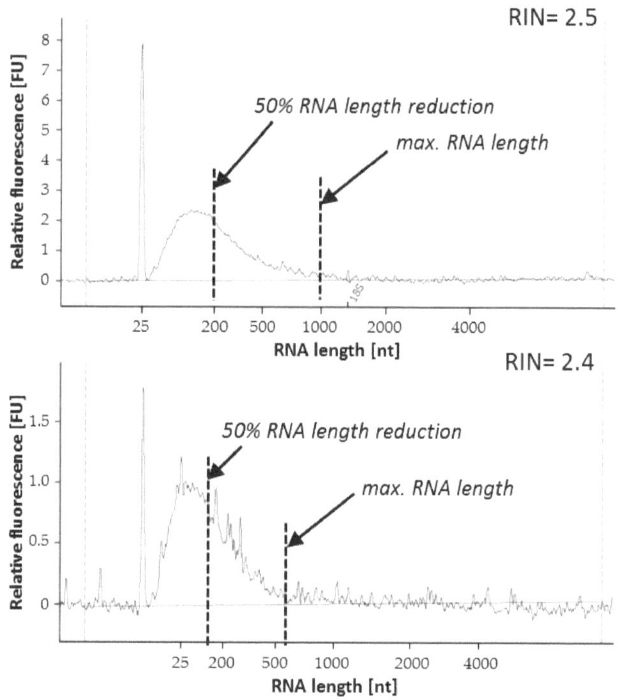

Figure 1. Evaluation of the extent of RNA fragmentation from FFPE tissue samples measuring the area under the curve from Agilent Bioanalyzer. Two values were defined: maximal RNA length and the 50% RNA length calculated as a 50% reduction of the area under the curve. RIN: RNA integrity number; FFPE: formalin-fixed paraffin-embedded specimens. nt: number of nucleotides; FU: fluorescence unit.

2.5. Analysis of RNA Quality from FFPE Biospecimens by Expression of Home-Keeping Genes

For PCR analysis, RNA was reverse-transcribed into complementary DNA (cDNA) using random hexamer primers and cDNA Synthesis Kit RevertAid (Fermentas, St. Leon-Rot, Germany). Quantitative qRT-PCR was performed using StepOnePlus RT-PCR-System (Applied Biosystems/Life Technologies, Darmstadt, Germany) and TaqMan primers for two housekeeping genes—glyceraldehyde 3-phosphate dehydrogenase (*GAPDH*, Hs04420697_g1, 130 bp) and beta-actin (*ACTB*, Hs01060665_g1, 63 bp)—to determine the gene expression level from various tissues samples over time between 2009 and 2018.

2.6. Statistical Analysis

For statistical analysis, SPSS for Windows version 20.0 (SPSS Inc., Chicago, IL, USA) was applied. Kruskal–Wallis test was used to set statistically significant differences between all study groups. The nonparametric Mann–Whitney *U* test was applied as a post-hoc comparison between the interindividual values. A p-value of <0.05 was considered as statistically significant.

3. Results

3.1. Study Population

Starting in January 2004 until December 2018, vascular tissue was collected from over 2700 patients (carotid plaques, $n = 1567$; AAA wall specimens, n = 481; PAD, $n = 703$, since 2016, including peripheral aneurysms, $n = 64$ and thrombus, $n = 80$) (Figure 2, Table 1). Furthermore, serum was collected from over 4400 patients (patients with high-grade carotid stenosis, $n = 1394$; AAA patients, $n = 1380$; PAD patients, $n = 1702$). For comparison, we also collected healthy aortic tissue samples as control from the Department of Trauma Surgery following kidney transplantation ($n = 102$). In addition, 15 healthy carotid arteries were obtained from the Department of Forensic Medicine.

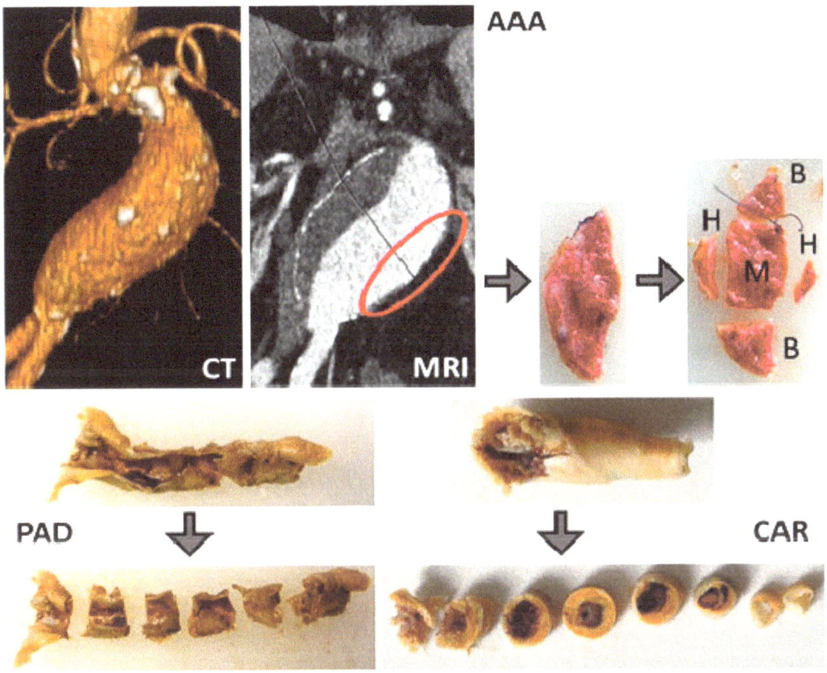

Figure 2. Examples of tissue samples collected in our Munich Vascular Biobank. CAR: carotid atherosclerotic plaques obtained from patients with high-grade carotid artery stenosis (>50%) [16] by endarterectomy (EA); PAD: atherosclerotic plaques from patients with peripheral artery disease obtained by EA [17]; AAA: aortic wall from patients with abdominal aortic aneurysm who underwent open surgical repair [18]; B: fresh frozen segments for molecular biology; H: histology; M: mechanics (tensile tests); CT: computer tomography, MRI: magnetic resonance imaging.

In addition, clinical data from all patients whose tissue was collected in our Munich Vascular Biobank were acquired (if available), including age, sex, medication, accompanying diseases, and findings from the Department of Clinical Chemistry. The collection of vascular tissue samples has proved its usefulness leading to plethora of scientific publications (Table 2) and helping to discover new features contributing to AAA, PAD and carotid plaque vulnerability.

Table 1. Munich Vascular Biobank. List of samples collected over the years.

Year	AAA #		CAROTIS		PAD	
	FFPE Tissue	Serum	FFPE Tissue	Serum	FFPE Tissue	Serum
2004			79	47		
2005	8	17	72	63		
2006	36	84	36	84		
2007	40	89	79	56		
2008	34	84	63	72		
2009	40	60	100	62	63	90
2010	40	72	94	58	77	78
2011	38	91	121	87	51	133
2012	33	136	122	111	62	251
2013	41	114	126	99	63	228
2014	24	131	92	97	63	259
2015	26	126	126	131	84	294
2016	30	124	124	117	80/27/35 *	228
2017	48	127	173	156	22/15/23 **	116
2018	43	125	160	154	16/22 **	25
Sum:	**481**	**1380**	**1567**	**1394**	**703**	**1702**

* Starting from 2016, we started to focus on peripheral aneurysm and thrombus as well (PAD/aneurysm/thrombus). ** Since 2017, we have focused only on peripheral aneurysm and thrombus (aneurysm/thrombus). # AAA tissue is increasingly difficult to obtain due to the fact that open aneurysm repair is more and more frequently replaced by endovascular techniques.

Table 2. List of publications resulting from our Munich Vascular Biobank.

Years	AAA	CAROTIS	PAD	Σ
2004–2008	1	1	1	3
2009–2013	17	20	3	40
2014–2018	10	23	6	41
Sum:	**29**	**44**	**11**	**84**

3.2. Tissue Processing and Characterization

For a better understanding of the development of atherosclerotic plaques or AAA, and to link these findings with patient medical history and accompanying diseases, a large number of specimens is necessary. Such endeavor requires good organization and logistics and is time-consuming. A particularly critical issue here is contemporary, accurate, and fast tissue processing. Furthermore, additional work is necessary, such as continuous plaque characterization and acquirement of available patient data, especially regarding the correct diagnosis.

Here, we provide an example of the tissue sample processing in our Munich Vascular Biobank (Figure 3):

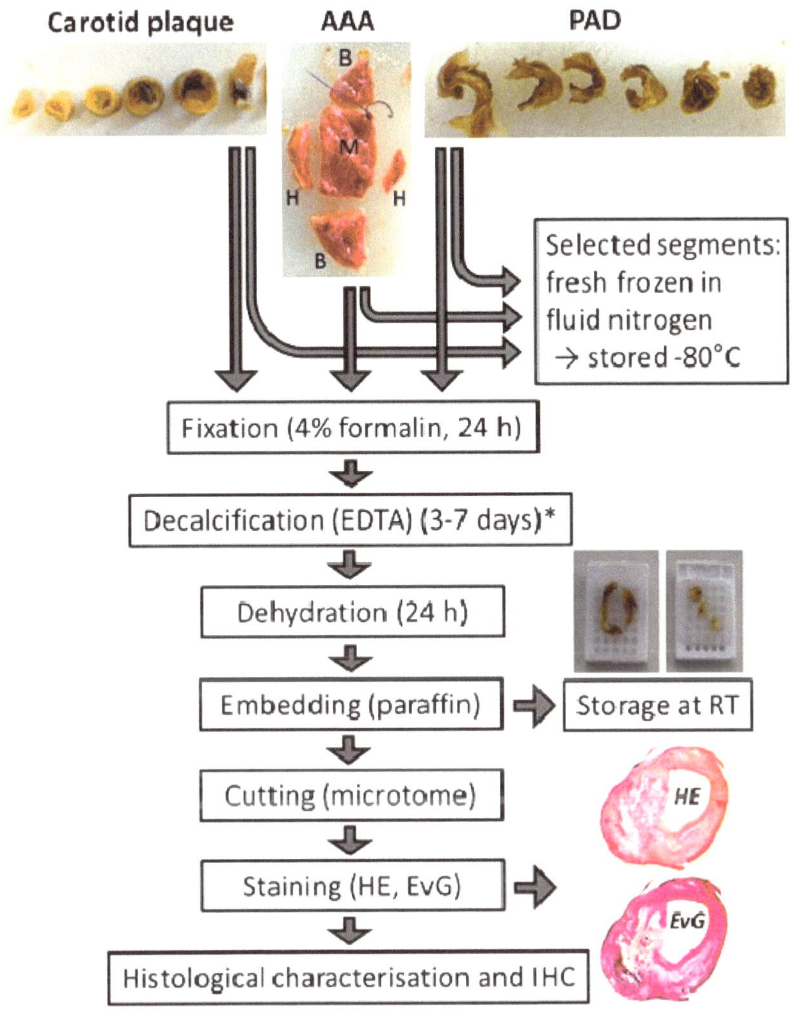

Figure 3. Schematic chart of the processing of vascular tissue after surgical excision. Carotid plaque: atherosclerotic lesions from patients with high-graded carotid artery stenosis; AAA: aortic wall from patients with abdominal aortic aneurysm; PAD: atherosclerotic tissue from patients with peripheral artery disease; RT: room temperature; IHC: immunohistochemistry. HE: haematoxilin-eosin staining; EvG: elastica van gieson staining; * the time depends on the extent of calcification.

Carotid artery tissue biospecimens: Following surgical excision, the complete atherosclerotic plaque is divided into segments of 3–4 mm each (mostly 3–7, depending on the size of the excised plaque tissue). Selected segments are immediately frozen and stored for molecular analyses. The remaining segments are fixed in formalin, decalcified, and embedded in paraffin as described in the Experimental Section. In this way, over 5000 carotid atherosclerotic plaque segments of different stages of atherosclerosis have been collected. From all segments, consecutive sections are continuously prepared, stained, and characterized to assess histological quality, cellular composition, the degree of infiltration with inflammatory cells, stage and extent of calcification, neovascularization, and the content of elastin and collagen fibers.

Furthermore, the atherosclerotic segments are classified in accordance with the AHA [24–27] to assess the type of atherosclerosis (Figure 4). In addition, plaque stability is determined in line with the criteria outlined by Redgrave and Rothwell [28], i.e., stable lesions are defined as plaques with a thick-cap fibroatheroma of >200 μm over a lipid/necrotic core (or without), while unstable lesions are defined either as ruptured or as plaques with thin-cap fibroatheroma of <200 μm over a large necrotic core.

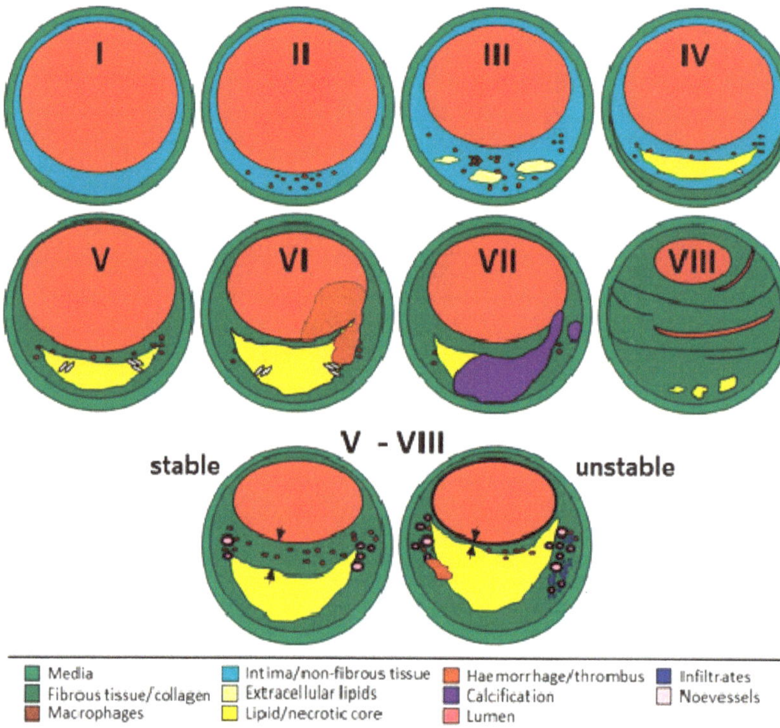

Figure 4. Classification of atherosclerotic lesions according to the American Heart Association (AHA) [24–27]. Type I: initial lesion with isolated macrophages and macrophage-derived foam cells; type II: fatty streaks, increased number of foam cells, intracellular lipid accumulation; type III: further accumulation of inflammatory cells and intracellular lipids, isolated extracellular lipid deposits; type IV: atheroma, formation of confluent lipid core without perceptible fibrous cap; type V: fibroatheroma, formation of fibrous layer over the lipid/necrotic core; type VI: as V but with thrombus and/or intraplaque hemorrhage; type VII: as V with calcified nodules, calcification predominates; type VIII: fibrous tissue predominates, lumen mainly small, lipid deposits minimal or absent. Plaque stability was assessed in line with [28]: thin fibrous cap <>200 μm (arrows) over a larger necrotic core. Unstable/vulnerable plaque can develop from each plaque type of type V–VIII. Modified from [29].

Peripheral artery tissue biospecimens: These tissue samples resemble carotid plaques [30] and are treated in a similar way as described above. However, plaque stability is not assessed, and plaque classification is performed only for selected tissue samples [30]. Furthermore, a majority of these biospecimens are FFPE-treated with a focus on histology.

AAA tissue biospecimens: Following surgical excision of a piece of the diseased aortic wall from the left anterior section of the aneurysm, the tissue is divided for histological and molecular biological analyses, as shown in Figure 3. A selected number of AAA wall samples also undergo tensile tests (M) to obtain additional information about the mechanical properties of the AAA wall [31–35].

Appointed segments (Figure 3) are immediately frozen in liquid nitrogen and stored accordingly (B). The remaining segments (H) are fixed in formalin, decalcified, embedded in paraffin, and stored at room temperature as described in the Experimental Section. Again, consecutive slices of 2–3 μm are prepared from all FFPE segments, stained, and characterized to assess tissue morphology, quality, and other histological features as described above.

3.3. *Analysis of RNA Quality from FFPE Biospecimens by RNA Integrity Number*

Because most of our vascular tissue samples are FFPE, in this work, we focused on the suitability of these biospecimens for expression analysis. The RNA quality stated by the RIN was unchanged for all time points for up to 10 years back, without any significant differences between the individual tested years (Figure 5A). The average RIN from all FFPE biospecimens analyzed in the study was 2.0 ± 0.6.

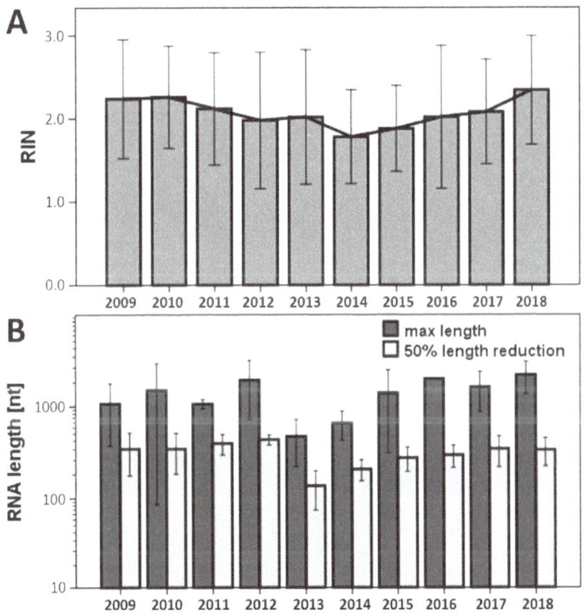

Figure 5. (**A**) Measurement of RNA integrity number (RIN) in FFPE vascular tissue samples using Agilent Bioanalyzer between 2009 and 2018 ($n = 5$ for each group and year). (**B**) Evaluation of the length of the RNA fragments, as described in Figure 3. No significant differences were observed between the study years over time.

3.4. *Analysis of RNA Quality from FFPE Biospecimens Evaluating RNA Fragmentation*

Independent of RIN, which is determined by comparing the degradation of 18S and 28S RNAs, the mRNA of interest can be differentially fragmented independently of the mentioned ribosomal RNAs. Thus, measuring the overall RNA fragmentation may provide additional information about the RNA quality and the maximal and average length (50% of the area under the curve from the Bioanalyzer, Figure 1) of RNA fragments in all tissue sample analyzed, as also partially described by Illumina [36]. The maximal length of the RNA of all vascular biospecimens exceeded 500 nt, while most of them were even longer than 1000 nt (Figure 5B). Calculating the drop-down of 50% showed that more than half of the RNA fragments in FFPE tissue samples was still longer than 200 nt. Furthermore, no significant differences in the overall RNA fragmentation were observed for up to 10 years back.

3.5. Analysis of RNA Quality from FFPE Biospecimens by Expression of Housekeeping Genes

Independent of RIN values or the degree of RNA degradation to assess the quality of RNA, expression analysis using various housekeeping genes is another important and helpful measure to prove the suitability of tissue samples for analyses, such as for expression analysis. Consequently, independent of the results from the Bioanalyzer (RIN, overall RNA fragmentation), we evaluated the expression of two housekeeping genes: *GAPDH* and *ACTB* (Figure 6). The mRNA expression from the selected FFPE vascular tissues was detected in all biospecimens. Using the same threshold for all PCR reactions, no significant differences were observed for cycles crossing the chosen threshold, independent of the selected year (up to 10 years back). The expression at mRNA level depended only on the concentration of the total amount of extracted RNA used in the study (data not shown). Furthermore, no significant relationship was observed between RIN, total RNA fragmentation, and mRNA expression.

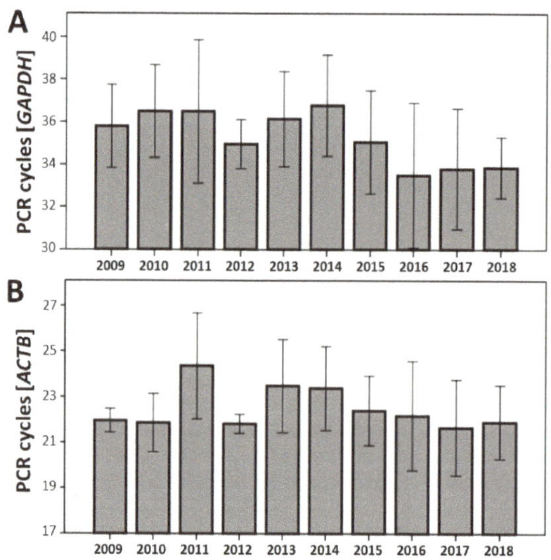

Figure 6. Results of qRT-PCR analysis from FFPE vascular tissue samples from different years between 2009 and 2018 (n = 5 for each study group) using TaqMan primer for glyceraldehyde 3-phosphate dehydrogenase (**A**, *GAPDH*, 130 bp) and beta-actin (**B**, *ACTB*, 63 bp). No significant differences were observed between the individual years over time.

4. Discussion

The high quality of biospecimens is an important factor for consistent and reliable analysis, and global harmonization of corresponding protocols is necessary to reduce variability between different biobanks [37–39]. The critical step is the time span between surgical excision and tissue preservation, which markedly influences, in particular, the quality of RNA in the acquired tissue samples [19–23]. In addition, to ensure the high quality of the biospecimens, suitable tests should be developed and applied to assess the quality of the biospecimens for the desired experiments. Unfortunately, a majority of publications dealing with tissue material from various biobanks do not cite how the biospecimens were obtained and how they were processed [40]. The most commonly used techniques for preservation of biomaterial are based on hypothermic or chemical form of treatment. Though both procedures are considered the most suitable methods for storage, they have their drawbacks, which might hamper the quality of the biospecimens and consequently the possible outcome of the results [41–43].

The RNA integrity is frequently used as a measure of tissue quality because RNA particularly rapidly degrades after tissue excision or after death due to its highly unstable nature. However, in contrast to the vulnerable purified RNA accessible to RNases, which is ubiquitously present, the RNA in tissue seems to be more stable [44]. A plethora of studies have already found that the RNA is stable within the tissue for several hours even at room temperature [22,44–46]. RNA was stable for up to three hours at RT in human breast biopsies [45], for up to five hours in human lung tissue [46], for up to 16 h on ice or at RT in human tonsil and colon tissue [44], and for up to 12 h at RT and 24 h on ice in liver tissue [22]. In addition, the RNA integrity is not significantly decreased after surgery [47]. Nevertheless, depending on the tissue samples stored in the individual biobanks and the time span between surgical excision and sample processing, the integrity of RNA should be tested beforehand to guarantee reliable results, even if such a verification is time-consuming and costly.

Many biobanks around the world are driven by institutes of pathology that commonly obtain tissue samples from different surgical departments to issue diagnostic findings and proper therapeutic strategies. Fixation, particularly in formalin, followed by embedding in paraffin (FFPE) is a very common technique to preserve human tissue samples. Consequently, FFPE tissue samples represent a great challenge in molecular biology because the fixation process causes nucleic acid degradation, resulting in fragmented RNA transcripts [48,49]. A common method to assess the integrity of RNA used by many researchers is the measurement of the RIN [50]. Our results demonstrated no significant differences in RIN for up to 10 years back, indicating that the RNA is stable over time even if the FFPE biospecimens are stored at room temperature. In this work, we also proposed an additional method to evaluate the RNA quality by assessing the degree of total RNA fragmentation. The determination of the maximal length of all the RNA fragments and 50% drop-down confirmed that no significant degradation of RNA in FFPE tissue samples occurred over time. Many of the biospecimens had maximal available RNA length over 1000 nt. Taking into account the 50% reduction of the RNA length, a majority of the tissue samples still contained RNA that were more than 200 nt long. Thus, in accordance with Illumina RNA enrichment assays and recommendations [36], these samples might also be suitable for RNA sequencing.

In order to further support our assumption that FFPE tissue samples are suitable for expression analyses at the mRNA level, we tested two different housekeeping genes: *GAPDH* and *ACTB*. Again, independent of the RIN value (even if lower than 1) or extended overall RNA fragmentation, expression of these genes could be detected in all FFPE biospecimens. Furthermore, no significant differences in the expression were observed over time, confirming the tissue stability and the fact that the storage of FFPE tissue samples at RT was adequate. No significant correlation was found between RIN values, global RNA fragmentation, and gene expression. The expression level depended only on the amount of RNA extracted from each sample. These results accredit that FFPE samples can be used without any concerns for expression analyses as well. However, the PCR amplicons should be kept short (<150 nt, preferably <100 nt) [51]. Furthermore, RIN measurement and RNA fragmentation analysis from the Bioanalyzer traces might be helpful in evaluating the quality of RNA and to exclude potential samples unsuitable for the intended experiments. In addition, due to the RNA degradation in FFPE tissue, hexamer primers should be used to prepare cDNA for RT-qPCR [22]. Using oligo (dT) primer for reverse transcription might lead to inconsistent results because poly A tail might be missing in many mRNA molecules, which also depends on the length and stability of the targeted mRNA. It should also be mentioned that the degradation of mRNA from various genes might differ from the housekeeping genes used in our study. Furthermore, genes expressed at high level, such as the housekeeping genes, may not work optimally to measure subtle differences in quality [52,53]. Thus, testing of various primer pairs from different mRNA regions for each individual transcript of interest is recommended to achieve reliable results.

In order to link the results from the vascular tissue to the patient's clinical data, histological features from all segments of each patient were compared to select the most clinically relevant characteristics. The following criteria and priorities were applied [29,54,55]: (i) plaque vulnerability (unstable > stable), (ii) plaque type (complex plaque > VI >V > VII), (iii) inflammation (positive > absent), and (iv) content of collagenous fibers (absent > positive). In this manner, we compared atherosclerotic plaques with sex, age, and history of neurological symptoms in 763 patients with high-grade carotid artery stenosis [54]. Male sex was significantly associated with increased inflammation and neovascularization. Higher age correlated with calcification, and unstable plaques were found more frequently in symptomatic patients. Interestingly, plaque morphology significantly differed between men and women and continuously changed with age, even though it was less striking than expected. Furthermore, our results showed that, independent of age, asymptomatic men had unstable plaques more frequently than women [55]. Thus, apart from age, male sex particularly seems to be an additional risk factor for ischemic stroke.

Based on our long-standing experience, each institute or research group intending to run a biobank should consider the following items before starting such a complex and ambitious endeavor: (i) A decision must be made on what tissue samples are to be collected and for what purpose. In this context, potential future perspectives and intentions should be considered, e.g., next-generation sequencing, omics analyses, etc. (ii) An organizational chart should be prepared in line with the corresponding surgical departments, medical specialists, pathologists, and/or biologists. (iii) The logistics is a critical item that should be considered to ensure the contemporary transfer of the biomaterial after tissue excision, its processing, and preservation. (iv) Standard protocols should be established for proper tissue processing, segmentation, and preservation techniques (cryopreservation, fixation with formalin, etc.). (v) Proper storage should be ensured for individual tissue samples at all times. FFPE can be stored at RT, as shown in our study. Fresh frozen tissue samples should be stored at a minimum of −80 °C; even better would be to store them at −150 °C (cryogenic freezer) or in liquid nitrogen tanks (−196 °C). (vi) Tissue collection alone does not make a good biobank; data acquisition from the patients, including medical history, accompanying diseases and other available datasets are important and helpful tools to link the results from tissue analyses to the corresponding diseases. (vii) Obtaining adequate control tissue samples is another critical point. Healthy individuals rarely undergo surgical intervention, so the access is very limited. (viii) Last but not least, ethical approval from the local ethics committee and permission from each patient to collect the tissue samples are necessary, along with the General Data Protection Regulation to guarantee patient privacy. All data should be appropriately protected, and only authorized persons should have access.

5. Conclusions

Biobanking is the most appropriate library of biological materials for not only scientific research but also for clinical usage. The accuracy of the data generated from the biobank strongly depends on the quality of the stored tissue samples. Thus, the major challenge is the standardization of protocols for biospecimen processing and preservation.

In this work, we summarized our own experiences in managing a vascular biobank, starting with tissue excision through different processing techniques, testing the quality of the biospecimens, up to their proper storage. Furthermore, we provided evidence that FFPE tissue samples are suitable not only for histological and immunohistochemical analyses but also for expression analyses at the mRNA level and potential RNA sequencing. In addition, apart from RIN, we offered an additional tool to analyze the quality of tissue samples by calculating the extent of the RNA fragmentation. Such an approach ensures accurate selection of suitable samples for the desired experimental design.

Biobank of the future is facing many challenges to be able to discover, develop, and properly validate new diagnostic and therapeutic strategies for basic, translational, and clinical research. Biobanking of high-quality human biospecimens, together with patient clinical information, provides a fundamental scientific infrastructure for personalized medicine.

Author Contributions: J.P. wrote and designed the manuscript; R.H. processed biological tissue samples and performed histological staining and histological tissue characterization; S.B., S.M., and J.P. participated in tissue characterization and collection of clinical data; N.G. performed experiments on RNA stability; A.B., B.R., M.K., and M.T. participated in the collection of clinical data and tissue samples; A.Z., H.W., P.T., S.S., C.K., C.S., T.S., G.B., U.W., F.M., K.S., A.-L.M., O.R., S.D.-S., C.R., and A.K. contributed by collecting tissue samples; E.K. contributed by organizing blood sample collection; A.B., L.M., A.Z., P.T., and H.-H.E. critically revised the manuscript. All authors read and approved the final version of the manuscript.

Acknowledgments: The Munich Vascular Biobank was supported by the Department of Vascular and Endovascular Surgery, Klinikum rechts der Isar der Technischen Universtität München, Germany. We also thank Doctor Edouard Matevossian from the Department of Surgery for providing us with the healthy aortic tissue samples.

Conflicts of Interest: The authors declare no conflict of interest.

References

1. De Souza, Y.G.; Greenspan, J.S. Biobanking past, present and future: Responsibilities and benefits. *AIDS* **2013**, *27*, 303. [CrossRef] [PubMed]
2. Diaz, Z.; Aguilar-Mahecha, A.; Paquet, E.R.; Basik, M.; Orain, M.; Camlioglu, E.; Constantin, A.; Benlimame, N.; Bachvarov, D.; Jannot, G.; et al. Next-generation biobanking of metastases to enable multidimensional molecular profiling in personalized medicine. *Mod. Pathol.* **2013**, *26*, 1413–1424. [CrossRef] [PubMed]
3. Busch, A.; Eken, S.M.; Maegdefessel, L. Prospective and therapeutic screening value of non-coding RNA as biomarkers in cardiovascular disease. *Ann. Transl. Med.* **2016**, *4*, 236. [CrossRef] [PubMed]
4. Kandpal, R.; Saviola, B.; Felton, J. The era of omics unlimited. *Biotechniques* **2009**, *46*, 351–355. [CrossRef] [PubMed]
5. Virchow, R. A more precise account of fatty metamorphosis. In *Cellular Pathology as Based Upon Physiological and Pathological History (English Translation of Second German Edition) Lecture XV: JP*; Lippincott: Philadelphia, PA, USA, 1971; pp. 350–366.
6. Ross, R. Atherosclerosis—An inflammatory disease. *N. Engl. J. Med.* **1999**, *340*, 115–126. [CrossRef] [PubMed]
7. Falk, E. Morphologic features of unstable atherothrombotic plaques underlying acute coronary syndromes. *Am. J. Cardiol.* **1989**, *63*, E114–E120. [CrossRef]
8. Davies, M.J.; Richardson, P.D.; Woolf, N.; Katz, D.R.; Mann, J. Risk of thrombosis in human atherosclerotic plaques: Role of extracellular lipid, macrophage, and smooth muscle cell content. *Br. Heart J.* **1993**, *69*, 377–381. [CrossRef]
9. Moreno, P.R.; Falk, E.; Palacios, I.F.; Newell, J.B.; Fuster, V.; Fallon, J.T. Macrophage infiltration in acute coronary syndromes. Implications for plaque rupture. *Circulation* **1994**, *90*, 775–778. [CrossRef]
10. Davies, M.J. Stability and instability: Two faces of coronary atherosclerosis. The Paul Dudley White Lecture 1995. *Circulation* **1996**, *94*, 2013–2020. [CrossRef]
11. Virmani, R.; Kolodgie, F.D.; Burke, A.P.; Farb, A.; Schwartz, S.M. Lessons from sudden coronary death: A comprehensive morphological classification scheme for atherosclerotic lesions. *Arterioscler. Thromb. Vasc. Biol.* **2000**, *20*, 1262–1275. [CrossRef]
12. Hurks, R.; Hoefer, I.E.; de Kleijn, D.P.; Daemen, M.J.; Moll, F.L.; Pasterkamp, G. Past, current and future concepts in atherosclerotic biobanking. *Future Cardiol.* **2008**, *4*, 639–649. [CrossRef] [PubMed]
13. Hellings, W.E.; Moll, F.L.; de Kleijn, D.P.; Pasterkamp, G. 10-years experience with the Athero-Express study. *Cardiovasc. Diagn. Ther.* **2012**, *2*, 63–73. [PubMed]
14. Redgrave, J.N.; Lovett, J.K.; Rothwell, P.M. Histological features of symptomatic carotid plaques in relation to age and smoking: The Oxford plaque study. *Stroke* **2010**, *41*, 2288–2294. [CrossRef] [PubMed]
15. Vähämurto, L.; Pahkala, K.; Magnussen, C.G.; Hutri-Kähönen, N.; Kähönen, M.; Laitinen, T.; Taittonen, L.; Tossavainen, P.; Lehtimäki, T.; Jokinen, E.; et al. Coronary heart disease risk factor levels in eastern and western Finland from 1980 to 2011 in the cardiovascular risk in Young Finns study. *Atherosclerosis* **2018**, *280*, 92–98. [CrossRef] [PubMed]
16. Abbott, A.L.; Paraskevas, K.I.; Kakkos, S.K.; Golledge, J.; Eckstein, H.H.; Diaz-Sandoval, L.J.; Cao, L.; Fu, Q.; Wijeratne, T.; Leung, T.W.; et al. Systematic Review of Guidelines for the Management of Asymptomatic and Symptomatic Carotid Stenosis. *Stroke* **2015**, *46*, 3288–3301. [CrossRef]
17. Espinola-Klein, C. ESC guidelines 2017 on peripheral arterial diseases: Summary of the most important recommendations and innovations. *Herz* **2017**, *42*, 721–727. [CrossRef] [PubMed]

18. Powell, J.T.; Ambler, G.K.; Svensjö, S.; Wanhainen, A.; Bown, M.J. Beyond the AAA Guidelines: Core Outcome Sets to Make Life Better for Patients. *Eur. J. Vasc. Endovasc. Surg.* **2019**, *57*, 6–7. [CrossRef] [PubMed]
19. Caixeiro, N.J.; Lai, K.; Lee, C.S. Quality assessment and preservation of RNA from biobank tissue specimens: A systematic review. *J. Clin. Pathol.* **2016**, *69*, 260–265. [CrossRef]
20. Ikeda, K.; Ichihara, K.; Hashiguchi, T.; Hidaka, Y.; Kang, D.; Maekawa, M.; Matsumoto, H.; Matsushita, K.; Okubo, S.; Tsuchiya, T.; et al. Evaluation of the short-term stability of specimens for clinical laboratory testing. Biopreserv. *Biobank.* **2015**, *13*, 135–143. [CrossRef]
21. Hubel, A.; Spindler, R.; Skubitz, A.P. Storage of human biospecimens: Selection of the optimal storage temperature. *Biopreserv. Biobank.* **2014**, *12*, 165–175. [CrossRef]
22. Lee, S.M.; Schelcher, C.; Gashi, S.; Schreiber, S.; Thasler, R.M.; Jauch, K.W.; Thasler, W.E. RNA stability in human liver: Comparison of different processing times, temperatures and methods. *Mol. Biotechnol.* **2013**, *53*, 1–8. [CrossRef] [PubMed]
23. Chevyreva, I.; Faull, R.L.; Green, C.R.; Nicholson, L.F. Assessing RNA quality in postmortem human brain tissue. *Exp. Mol. Pathol.* **2008**, *84*, 71–77. [CrossRef] [PubMed]
24. Stary, H.C. Natural history and histological classification of atherosclerotic lesions: An update. *Arterioscler. Thromb. Vasc. Biol.* **2000**, *20*, 1177–1178. [CrossRef] [PubMed]
25. Stary, H.C.; Blankenhorn, D.H.; Chandler, A.B.; Glagov, S.; Insull, W., Jr.; Richardson, M.; Rosenfeld, M.E.; Schaffer, S.A.; Schwartz, C.J.; Wagner, W.D.; et al. A definition of the intima of human arteries and of its atherosclerosis-prone regions. A report from the Committee on Vascular Lesions of the Council on Arteriosclerosis, American Heart Association. *Arterioscler. Thromb.* **1992**, *12*, 120–134. [CrossRef] [PubMed]
26. Stary, H.C.; Chandler, A.B.; Dinsmore, R.E.; Fuster, V.; Glagov, S.; Insull, W., Jr.; Rosenfeld, M.E.; Schwartz, C.J.; Wagner, W.D.; Wissler, R.W.; et al. A definition of advanced types of atherosclerotic lesions and a histological classification of atherosclerosis. A report from the Committee on Vascular Lesions of the Council on Arteriosclerosis, American Heart Association. *Circulation* **1995**, *92*, 1355–1374. [CrossRef] [PubMed]
27. Stary, H.C.; Chandler, A.B.; Glagov, S.; Guyton, J.R.; Insull, W., Jr.; Rosenfeld, M.E.; Schaffer, S.A.; Schwartz, C.J.; Wagner, W.D.; Wissler, R.W. A definition of initial, fatty streak, and intermediate lesions of atherosclerosis. A report from the Committee on Vascular Lesions of the Council on Arteriosclerosis, AHA. *Arterioscler. Thromb.* **1994**, *14*, 840–856. [CrossRef] [PubMed]
28. Redgrave, J.N.; Gallagher, P.; Lovett, J.K.; Rothwell, P.M. Critical cap thickness and rupture in symptomatic carotid plaques: The oxford plaque study. *Stroke* **2008**, *39*, 1722–1729. [CrossRef]
29. Pelisek, J.; Wendorff, H.; Wendorff, C.; Kuehnl, A.; Eckstein, H.H. Age-associated changes in human carotid atherosclerotic plaques. *Ann. Med.* **2016**, *48*, 541–551. [CrossRef]
30. Zimmermann, A.; Senner, S.; Eckstein, H.H.; Pelisek, J. Histomorphological evaluation of atherosclerotic lesions in patients with peripheral artery occlusive disease. *Adv. Med. Sci.* **2015**, *60*, 236–239. [CrossRef]
31. Gee, M.W.; Reeps, C.; Eckstein, H.H.; Wall, W.A. Prestressing in finite deformation abdominal aortic aneurysm simulation. *J. Biomech.* **2009**, *42*, 1732–1739. [CrossRef]
32. Maier, A.; Gee, M.W.; Reeps, C.; Eckstein, H.H.; Wall, W.A. Impact of calcifications on patient-specific wall stress analysis of abdominal aortic aneurysms. *Biomech. Model Mechanobiol.* **2010**, *9*, 511–521. [CrossRef] [PubMed]
33. Reeps, C.; Gee, M.W.; Maier, A.; Gurdan, M.; Eckstein, H.H.; Wall, W.A. The impact of model assumptions on results of computational mechanics in abdominal aortic aneurysm. *J. Vasc. Surg.* **2010**, *51*, 679–688. [CrossRef] [PubMed]
34. Maier, A.; Essler, M.; Gee, M.W.; Eckstein, H.H.; Wall, W.A.; Reeps, C. Correlation of biomechanics to tissue reaction in aortic aneurysms assessed by finite elements and [18F]-fluorodeoxyglucose-PET/CT. *Int. J. Numer. Method Biomed. Eng.* **2012**, *28*, 456–471. [CrossRef] [PubMed]
35. Reeps, C.; Maier, A.; Pelisek, J.; Härtl, F.; Grabher-Meier, V.; Wall, W.A.; Essler, M.; Eckstein, H.H.; Gee, M.W. Measuring and modeling patient-specific distributions of material properties in abdominal aortic aneurysm wall. *Biomech. Model Mechanobiol.* **2013**, *12*, 717–733. [CrossRef] [PubMed]
36. Evaluating RNA Quality from FFPE Samples. Available online: https://emea.illumina.com/content/dam/illumina-marketing/documents/products/technotes/evaluating-rna-quality-from-ffpe-samples-technical-note-470-2014-001.pdf (accessed on 26 October 2016).

37. Moore, H.M.; Compton, C.C.; Alper, J.; Vaught, J.B. International approaches to advancing biospecimen science. *Cancer Epidemiol. Biomark. Prev.* **2011**, *20*, 729–732. [CrossRef] [PubMed]
38. Moore, H.M.; Kelly, A.B.; Jewell, S.D.; McShane, L.M.; Clark, D.P.; Greenspan, R.; Hayes, D.F.; Hainaut, P.; Kim, P.; Mansfield, E.A.; et al. Biospecimen reporting for improved study quality (BRISQ). *Cancer Cytopathol.* **2011**, *119*, 92–101. [CrossRef] [PubMed]
39. Malm, J.; Fehniger, T.E.; Danmyr, P.; Végvári, A.; Welinder, C.; Lindberg, H.; Appelqvist, R.; Sjödin, K.; Wieslander, E.; Laurell, T.; et al. Developments in biobanking workflow standardization providing sample integrity and stability. *J. Proteom.* **2013**, *95*, 38–45. [CrossRef] [PubMed]
40. Simeon-Dubach, D.; Perren, A. Better provenance for biobank samples. *Nature* **2011**, *475*, 454–455. [CrossRef]
41. Betsou, F.; Rimm, D.L.; Watson, P.H.; Womack, C.; Hubel, A.; Coleman, R.A.; Horn, L.; Terry, S.F.; Zeps, N.; Clark, B.J.; et al. What are the biggest challenges and opportunities for biorepositories in the next three to five years? *Biopreserv. Biobank.* **2010**, *8*, 81–88. [CrossRef]
42. Karimi-Busheri, F.; Zadorozhny, V.; Shawler, D.L.; Fakhrai, H. The stability of breast cancer progenitor cells during cryopreservation: Maintenance of proliferation, self-renewal, and senescence characteristics. *Cryobiology* **2010**, *60*, 308–314. [CrossRef]
43. Karimi-Busheri, F.; Zadorozhny, V.; Carrier, E.; Fakhrai, H. Molecular integrity and global gene expression of breast and lung cancer stem cells under longterm storage and recovery. *Cell Tissue Bank.* **2013**, *14*, 175–186. [CrossRef] [PubMed]
44. Micke, P.; Ohshima, M.; Tahmasebpoor, S.; Ren, Z.P.; Ostman, A.; Ponten, F.; Botling, J. Biobanking of fresh frozen tissue: RNA is stable in nonfixed surgical specimens. *Lab. Investig.* **2006**, *86*, 202–211. [CrossRef] [PubMed]
45. Ohashi, Y.; Creek, K.E.; Pirisi, L.; Kalus, R.; Young, S.R. RNA degradation in human breast tissue after surgical removal: A time-course study. *Exp. Mol. Pathol.* **2004**, *77*, 98–103. [CrossRef] [PubMed]
46. Jewell, S.D.; Srinivasan, M.; McCart, L.M.; Williams, N.; Grizzle, W.H.; LiVolsi, V.; MacLennan, G.; Sedak, D.D. Analysis of the molecular quality of human tissues: An experience from the Cooperative Human Tissue Network. *Am. J. Clin. Pathol.* **2002**, *118*, 733–741. [CrossRef] [PubMed]
47. Almeida, A.; Paul Thiery, J.; Magdelenat, H.; Radvanyi, F. Gene expression analysis by real-time reverse transcription polymerase chain reaction: Influence of tissue handling. *Anal. Biochem.* **2004**, *328*, 101–108. [CrossRef] [PubMed]
48. Von Ahlfen, S.; Missel, A.; Bendrat, K.; Schlimpberger, M. Determinants of RNA quality from FFPE samples. *PLoS ONE* **2007**, *2*, e1261. [CrossRef] [PubMed]
49. Penland, S.K.; Keku, T.O.; Torrice, C.; He, X.; Krishnamurthy, J.; Hoadley, K.A.; Woosley, J.T.; Thomas, N.E.; Perou, C.M.; Sandler, R.S.; et al. RNA expression analysis of formalin-fixed paraffin-embedded tumors. *Lab. Investig.* **2007**, *87*, 383–391. [CrossRef]
50. Agilent Technologies. RNA Integrity Number (RIN)—Standardization of RNA Quality Control. Publication PN 5989-1165EN. Available online: https://www.agilent.com (accessed on 21 January 2016).
51. Arzt, L.; Kothmaier, H.; Quehenberger, F.; Halbwedl, I.; Wagner, K.; Maierhofer, T.; Popper, H.H. Evaluation of formalin-free tissue fix. for RNA and microRNA studies. *Exp. Mol. Pathol.* **2011**, *91*, 490–495. [CrossRef]
52. Zhang, X.; Han, Q.Y.; Zhao, Z.S.; Zhang, J.G.; Zhou, W.J.; Lin, A. Biobanking of Fresh-Frozen Gastric Cancer Tissues: Impact of Long-Term Storage and Clinicopathological Variables on RNA Quality. *Biopreserv Biobank.* **2018**. [CrossRef]
53. Olivieri, E.H.; de Andrade Franco, L.; Pereira, R.G.; Mota, L.D.; Campos, A.H.; Carraro, D.M. Biobanking practice: RNA storage at low concentration affects integrity. *Biopreserv. Biobank.* **2014**, *12*, 46–52. [CrossRef]
54. Wendorff, C.; Wendorff, H.; Pelisek, J.; Tsantilas, P.; Zimmermann, A.; Zernecke, A.; Kuehnl, A.; Eckstein, H.H. Carotid plaque morphology is significantly associated with sex, age, and history of neurological symptoms. *Stroke* **2015**, *46*, 3213–3219. [CrossRef] [PubMed]
55. Wendorff, C.; Wendorff, H.; Kuehnl, A.; Tsantilas, P.; Kallmayer, M.; Eckstein, H.H.; Pelisek, J. Impact of sex and age on carotid plaque instability in asymptomatic patients-results from the Munich Vascular Biobank. *Vasa* **2016**, *45*, 411–416. [CrossRef] [PubMed]

Journal of
Clinical Medicine

MDPI

Article

Surface Modification of Electrospun Scaffolds for Endothelialization of Tissue-Engineered Vascular Grafts Using Human Cord Blood-Derived Endothelial Cells

Diana Catalina Ardila [1], Jr-Jiun Liou [1], David Maestas [2], Marvin J. Slepian [3,4,5,6], Michael Badowski [7], William R. Wagner [1,8,9], David Harris [7,10] and Jonathan P. Vande Geest [1,8,11,*]

1 Department of Bioengineering, University of Pittsburgh, Pittsburgh, PA 15219, USA; dca13@pitt.edu (D.C.A.); jrl101@pitt.edu (J.-J.L.); wagnerwr@upmc.edu (W.W.)
2 Department of Biomedical Engineering, Johns Hopkins University, Baltimore, MD 21231, USA; dmaesta1@jhmi.edu
3 Sarver Heart Center, The University of Arizona, Tucson, AZ 85721, USA; slepian@email.arizona.edu
4 The Arizona Center for Accelerated BioMedical Innovation, University of Arizona, Tucson, AZ 85721, USA
5 BIO5 Institute for Biocollaborative Research, The University of Arizona, Tucson, AZ 85721, USA
6 Interventional Cardiology, University of Arizona, Tucson, AZ 85721, USA
7 Arizona Health Science Center Biorepository, University of Arizona, Tucson, AZ 85724, USA; badowski@email.arizona.edu (M.B.); davidh@email.arizona.edu (D.H.)
8 McGowan Institute for Regenerative Medicine, University of Pittsburgh, Pittsburgh, PA 15219, USA
9 Department of Surgery, University of Pittsburgh, Pittsburgh, PA 15219, USA
10 Department of Immunobiology, Arizona Health Science Center Biorepository, University of Arizona, Tucson, AZ 85724, USA
11 Vascular Medicine Institute, University of Pittsburgh, Pittsburgh, PA 15219, USA
* Correspondence: jpv20@pitt.edu; Tel.: +1-412-624-6496

Received: 12 December 2018; Accepted: 1 February 2019; Published: 4 February 2019

Abstract: Tissue engineering has gained attention as an alternative approach for developing small diameter tissue-engineered vascular grafts intended for bypass surgery, as an option to treat coronary heart disease. To promote the formation of a healthy endothelial cell monolayer in the lumen of the graft, polycaprolactone/gelatin/fibrinogen scaffolds were developed, and the surface was modified using thermoforming and coating with collagen IV and fibronectin. Human cord blood-derived endothelial cells (hCB-ECs) were seeded onto the scaffolds and the important characteristics of a healthy endothelial cell layer were evaluated under static conditions using human umbilical vein endothelial cells as a control. We found that polycaprolactone/gelatin/fibrinogen scaffolds that were thermoformed and coated are the most suitable for endothelial cell growth. hCB-ECs can proliferate, produce endothelial nitric oxide synthase, respond to interleukin 1 beta, and reduce platelet deposition.

Keywords: Vascular tissue engineering; umbilical cord blood; human cord blood-derived endothelial cells; endothelialization; vascular graft

1. Introduction

Coronary heart disease (CHD) resulting from atherosclerosis remains the leading cause of death in the United States [1]. CHD alone caused approximately 1 of 7 deaths in 2013, and 660,000 hospitalizations as a result of myocardial infarctions [2]. Once a coronary artery is compromised, a vascular bypass is an option to restore blood flow to tissues distal of the restriction or blockage [3].

Artery bypass graft surgery procedures usually involve the replacement of a coronary artery with an autologous vessel such as the saphenous vein or internal mammary artery [4,5]. Nevertheless, autologous grafts are not always available due to preexisting vascular conditions, or their use in a prior bypass operation [6]. The commercial alternatives to autologous grafts are vessels made of synthetic materials such as polytetrafluoroethylene (PTFE) or polyethylene terephthalate (known as Dacron). These grafts have been implemented with some success in medium and large diameter vessel replacement, but their efficacy is limited when used in small diameter vessels (<6 mm) where low blood flow makes the synthetic graft difficult to cellularize and more prone to thrombus formation, calcification, and intimal hyperplasia [3,6,7]. Therefore, there is an urgent need for small diameter graft alternatives that are able to support cell growth and match the mechanical properties of a native coronary artery while reducing the risk of acute thrombus formation and restenosis [8,9].

Recently, tissue engineering has gained attention as an alternative approach for developing small diameter vascular grafts from biocompatible natural or synthetic polymers [10,11]. Different methods have been used to fabricate tissue-engineered vascular grafts (TEVGs), such as solvent casting, phase separation, and electrospinning [10]. Electrospinning, in particular, has been extensively utilized to create fibrous scaffolds produced from nonwoven meshes containing fibers with diameters from micrometers to nanometers [10,12–14]. Its ability to combine the mechanical durability of synthetic materials with the cell compatibility of natural polymers makes electrospinning particularly attractive for TEVGs [10]. Some studies have reported the application of polycaprolactone scaffolds for the fabrication of TEVGs [15–17]. Previously we found that the incorporation of both gelatin and fibrinogen enhances the compliance which is critical for a functional vascular graft [18]. To further enhance the biocompatibility of a scaffold, surface modification has been shown to promote cell attachment, viability, and biological response of cells in the biomaterial. One of the common surface modification techniques is extracellular matrix protein coating [11].

For a TEVG to be successful, it needs to promote the formation of a healthy endothelial cell monolayer in the lumen of the graft [19]. Endothelial cells play a critical role in the control of vascular function. They participate in all aspects of vascular homeostasis, but they are also critical in physiological or pathological processes like regulation of vascular tone, inflammation, and in the prevention of thrombosis and intimal hyperplasia [20–22]. Therefore, an establishment of a healthy endothelium is crucial for the long-term success of a TEVG [19]. Endothelial progenitor cells as a source of cells for TEVG may be obtained from many sources including bone marrow, peripheral blood, and cord blood. Cord blood is an ideal source, since it has not been exposed to exogenous conditions that could diminish its numbers or function, such as diabetes, autoimmune disease, or old age [23]. In recent years, human umbilical cord blood-derived endothelial cells (hCB-ECs) have been increasingly used for tissue engineering applications, due to the highly proliferative capacity of these stem and differentiated cells in vitro [24]. An increased rate of proliferation results in less time and costs involved when "seeding" the grafts. A variety of studies have shown that hCB-ECs develop into a homogeneous population of endothelial cells that can be passaged 50–100 times before reaching senescence or losing the differentiated endothelial cell phenotype [24–26] as compared to other sources of endothelial precursor cells (such as adult peripheral blood), owing in part to the "young" age of the CB cells, which again decreases overall cost. In addition, a younger source of stem cells and EC derived from those stem cells is advantageous with respect to potential negative epigenetic effects observed in adult sources of cells. Studies comparing the protein profile of hCB-ECs with human umbilical vein endothelial cells (HUVECs) concluded that hCB-ECs display a higher proliferative capacity, higher sensitivity to angiogenic factors, and a differentiated production of the antioxidant enzyme manganese superoxide dismutase, which makes hCB-ECs more tolerant to oxidative stress than HUVECs [24,27–32] and more likely to survive in vivo.

In this study, gelatin/fibrinogen/polycaprolactone scaffolds were electrospun and surface-modified through a thermoforming process and coated with a blend of collagen IV and fibronectin. hCB-ECs were tested for their ability to proliferate and remain on the surface of the scaffold in formation of a monolayer. We also assessed whether the attached cells could function similar to a mature endothelial cell layer quantifying their response to pro-inflammatory cytokine interleukin 1 beta (IL-1β), the production of endothelial nitric oxide synthase (eNOS), and antithrombotic capacities as compared to HUVECs. This study evaluates the suitability of a surface modified biomaterial for endothelialization of a vascular graft as well as the use of hCB-ECs as a cell source for cardiovascular tissue engineering applications.

2. Materials and Methods

2.1. Cell Isolation and Cell Characterization

Human umbilical vein endothelial cells (HUVECs) (ATCC, Manassas, VA, USA) were purchased and cultured per manufacturer's instructions as a positive control. Human cord blood-derived endothelial cells (hCB-ECs) were isolated as follows and de-identified umbilical cord blood (n = 3 donors) was obtained from the University of Arizona Biorepository per protocols approved by the University of Arizona's Institutional Review Board (IRB). After collection, the isolation and differentiation procedures followed those described by Javed et al. (2008) [33], with minor modifications. Briefly, cord blood (20–100 mL) was diluted 1:1 with Hank's balanced salt solution (HBSS), and then overlaid onto an equivalent volume of Histopaque 1077 (Sigma-Aldrich, St. Louis, MO, USA). To isolate mononuclear cells, the diluted cord blood was centrifuged for 30 min at room temperature at 740× g. The isolated mononuclear cells were washed and resuspended in 12 mL complete EGM-2 plus medium (Lonza, Basel, Switzerland), supplemented with 15% fetal bovine serum (Thermo Fisher Scientific, Pittsburgh, PA, USA). The cells were seeded onto 3 wells of a 6-well plate precoated with rat tail collagen type I (Life Technologies, Carlsbad, CA, USA) and maintained in a humidified environment at 37 °C and 5% CO_2. The medium was changed daily for the first 7 days and every other day until the first passage. Colonies of endothelial cells appeared between 5 and 22 days of culture.

Cell identity was confirmed by flow cytometry and immunocytochemistry (ICC) as previously described [31] using mouse anti-human primary antibodies (BD Biosciences, Franklin Lakes, NJ, USA). For flow cytometry, fluorescein isothiocyanate (FITC)-conjugated CD31 antibodies, allophycocyanin (APC)-conjugated CD105 antibodies, and phycoerythrin (PE)-conjugated CD45 antibodies were used. Fluorescence intensity per cell produced by the bound antibodies was measured and cells were counted using the LSRII flow cytometer (BD Biosciences, Franklin Lakes, NJ, USA). For each flow cytometry experiment, 10,000–30,000 cells were gated from a population of 500,000 cells. For ICC, cells were cultured on glass coverslips coated with rat tail collagen I and the immunostaining was performed using the FITC conjugated CD31 antibodies. Cell nuclei were counterstained using VECTSHIELD® 4′,6-diamidino-2-phenylindole (DAPI)-containing mounting media (Vector Laboratories, Burlingame, CA, USA). For all experiments performed, cells from passage 2–6 were used. Detailed information on the antibodies used is provided in Table S1.

2.2. Scaffold Fabrication

Flat sheet scaffolds were fabricated by electrospinning. Briefly, polycaprolactone, (PCL, 80,000 MW; Sigma-Aldrich, St. Louis, MO, USA), gelatin extracted from porcine skin (Sigma-Aldrich, St. Louis, MO, USA), and fraction I bovine fibrinogen (Sigma-Aldrich, St. Louis, MO, USA) were mixed at a ratio of 50% PCL:40% gelatin:10% fibrinogen w/w (hereafter PCL-GF) [18,34,35]. The polymeric blend was dissolved in 1,1,1,3,3,3-Hexafluoro-2-propanol (Sigma-Aldrich, St. Louis, MO, USA) to create a 10% w/v solution under constant stirring until completely homogeneous. The solution was loaded into a 5 mL BD syringe with a 23-gauge stainless steel dispensing blunt tip needle attached (CML Supply, Lexington, KY, USA). The syringe was then loaded into a NE-1000 single syringe pump (New Era Pump Systems Inc, Farmingdale, NY, USA) set to a pump rate of 30 μL/min. The distance from the needle tip to the target was 8 cm. The polymeric solution was electrospun with an applied voltage of 15 kV, onto glass coverslips attached to a metallic target to create fine fibers. The resulting flat sheets were cross-linked in 25% glutaraldehyde (Sigma-Aldrich, St. Louis, MO, USA) in vapor phase for 24 h. The glutaraldehyde was evaporated in a convection oven overnight at 42 °C, which is a temperature below the denaturation point of gelatin and fibrinogen [36,37]. Additionally, the scaffolds were rinsed with deionized water to remove any cross-linker residues and uncross-linked gelatin [35]. This method of removing glutaraldehyde was previously demonstrated no cytotoxicity of the cross-linked scaffolds for the proliferation of smooth muscle cells [35].

2.3. Surface Modification

After cross-linking, the scaffolds were thermoformed by a technique previously reported by our group [38]. The scaffolds were immersed in a water bath at 45 °C for 2 min, and then quickly placed in between two glass slides. A pressure of approximately 45 mmHg was applied using a 25 mm wide binder clip while the scaffolds were immersed once more in the 45 °C water bath for 5 min [39]. The scaffolds were equilibrated at room temperature for 10 min before the pressure was released.

Thermoformed scaffolds were sterilized with 70% ethanol for 1 h, rinsed with sterile 1× PBS and then placed under UV light (254 nm) for 1 h. The sterile scaffolds were coated with a solution 1:1 of collagen IV (Sigma-Aldrich, St. Louis, MO, USA) and fibronectin (Sigma-Aldrich, St. Louis, MO, USA) in HBSS with a final concentration of 5 μg/mL for 24 h at 4 °C. This coating blend has previously shown to promote hCB-EC growth when compared to Collagen IV alone (Figure S1). The coating solution was carefully rinsed with sterile 1× PBS. The scaffolds were then designated as thermoformed and coated (hereafter TC). The combined surface modification with thermoforming and coating was selected because our preliminary data showed that scaffolds thermoformed and coated are most favorable for cell attachment and spreading when compared to thermoformed alone or coated alone (Figure S2).

2.4. Scaffold Imaging

PCL-GF scaffolds, both nontreated (hereafter NT) and thermoformed, were imaged using atomic force microscopy (AFM). All AFM data was collected with a Cypher (Asylum Research and Oxford Instruments company, Santa Barbara, CA, USA) using AC Topography mode in air and NSC15 tapping mode probes (Mikromasch, Watsonville, CA, USA). For each scaffold treatment, 3 samples were imaged in 4 random areas of 25 μm^2 along the mesh of fibers, and roughness values (Ra) were calculated after applying a first order flatten on all surfaces using Asylum Research software version 14.13.134.

Scanning electron microscopy (SEM) was performed on both TC and NT scaffolds. Briefly, samples were mounted onto aluminum stubs, grounded with silver paint, and sputter-coated with 6 nm gold/palladium (Cressington Sputter-coater Auto 108, Cressington, Watford, UK). En face views of the samples (n = 6 scaffolds each) were imaged using a JEOL JSM-6335F scanning electron microscope (JEOL USA, Peabody, MA, USA) at 3 kV at a magnification of 10,000×. The SEM images were binarized and the porosity was calculated as the ratio of the total number of fiber pixels to the total number of pixels in the image. The fiber diameter was calculated by manually measuring the diameter of 120 randomly selected fibers per scaffold treatment via freehand lines superimposed over the SEM images in ImageJ.

Multiphoton microscopy was used for the three-dimensional imaging of our electrospun scaffolds. The Pitt Advanced Intravital Microscope (AIM) for multiphoton imaging at the University of Pittsburgh Soft Tissue Biomechanics Laboratory allowed us to measure the change in scaffold thickness. This Olympus BX51 upright laser scanning microscope (Olympus, Tokyo, Japan) was coupled to a 120-fs tunable pulsed Titanium-Sapphire laser (Coherent Inc, Santa Clara, CA, USA) and an Olympus XLUMPLFL 20× water immersion objective with a numerical aperture of 0.9 [40,41]. The fibers were imaged centering the laser at 780 nm to excite the autofluorescence signal from the scaffolds (NADH), split with a 568 nm dichroic mirror, and collected through a 525/50 nm bandpass filter. The signal was collected over a 400 μm × 400 μm field of view at 2-μm z-step-size along the scaffold thickness.

2.5. Effect of Surface Modification on Cell Growth

hCB-ECs and HUVECs were seeded in NT and TC scaffolds at 10,000 cells/scaffold and cultured for 7 days. The culture medium was changed every other day and cultures were maintained in a humidified environment at 37 °C and 5% CO_2. Cell growth was evaluated after 7 days of culture. A sample of approximately 25 mm^2 was cut from each scaffold, and cell number was measured by MTS assay. Briefly, cell-seeded scaffolds were incubated in culture medium supplemented with CellTiter 96 AQueous One Solution Cell Proliferation Assay (Promega, Madison, WI, USA) at 37 °C for 4 h. Supernatant was collected and the absorbance at 490 nm was recorded. Background absorbance from the NT and TC scaffolds was obtained from nonseeded scaffolds. Cell number was calculated based on our calibration curves (Figure S3).

For cell imaging, the scaffolds were fixed with 2% formaldehyde and stained with Alexa Fluor® 568 phalloidin (Life Technologies, Carlsbad, CA, USA) to visualize f-actin following the manufacturer's instructions. To stain the nuclei, the scaffolds were treated for 24 h with VECTASHIELD® DAPI mounting medium (Vector Laboratories, Burlingame, CA, USA). The Pitt AIM with a 20× water immersion objective was used to visualize the cells growing in the scaffolds along the scaffold depth. The nuclei (blue), fibers (green), and f-actin (red) were imaged simultaneously and colocalized using three different photomultiplier tubes (PMTs). The laser was centered at λ = 780 nm to excite simultaneously DAPI, the autofluorescence signal from the scaffolds (NADH), and Alexa Fluor® 568. In the first PMT, the DAPI signal was split with a 505 nm dichroic mirror and collected through a 460/80 bandpass filter. In the second PMT, the signal from the scaffolds was split with a 568 nm dichroic mirror and collected through a 525/50 bandpass filter. Alexa Fluor® 568 signal was acquired in the third PMT by splitting the signal with a 568 nm dichroic mirror and collecting using a 607/70 bandpass filter. The colocalized image stacks from the cell nuclei, the fibers, and f-actin were merged to visualize the cell location in the scaffolds. Maximum intensity projections (MIPs) were produced to visualize the total number of cells in the field of view. The percentage of cell infiltration was calculated as the ratio of the length that cells migrating through the flat sheet from the top to the bottom relative to the flat sheet thickness. For instance, in a cell-seeded flat sheet, we obtained 20 images every 2 µm in the Z-direction; the flat sheet had an estimated thickness of 40 µm. If the cells only appear in the first five images, it means the cells migrated through the first 2 × 5 = 10 µm. Therefore, the percentage of cell infiltration is calculated as 10/40 = 25% of that specific sample.

2.6. Platelet Activation and Platelet Adherence

De-identified human whole blood was obtained at the University of Arizona Sarver Heart Center in accordance with IRB-approved protocols. Briefly, 30 mL of whole blood was drawn through venipuncture into 3 mL of acid citrate dextrose A. Subsequently, the blood was centrifuged at 500× *g* for 15 min to obtain platelet-rich plasma, which was gel-filtered through a column of Sepharose 2B beads (GE Healthcare Life Sciences, Marlborough, MA, USA) with HEPES-modified Tyrode's buffer (Boston BioProducts Inc, Ashland, MA, USA), to collect erythrocyte-free platelets. With the aim of quantifying the activation of platelets caused by the contact with cell-seeded scaffolds, the platelet concentration was diluted to 20,000 platelets/µL in HEPES-modified Tyrode's buffer, with 3 mM $CaCl_2$ added 10 min prior to any experiment to avoid platelet-platelet activation.

Platelet activity state (PAS) was measured using a modified-prothrombinase assay reported in the literature [42,43]. Briefly, after 7 days of hCB-EC and HUVEC culture in NT or TC scaffolds, a sample of 20,000 platelets/µL was perfused onto the surface of the cell-seeded scaffolds and incubated at 37 °C for 0, 1, and 2 h. After incubation, 100 pM factor Xa, 5 mM calcium, and 200 nM acetylated prothrombin, were added and incubated for 10 min. Thrombin generation was quantified through spectrophotometric analysis over 7 min at an absorbance wavelength of 405 nm to obtain the PAS values, using Chromozym-TH (Roche, Penzberg, Germany) as the thrombin-specific chromogenic peptide substrate. The platelets in a control tube were fully activated by sonication. Each PAS value was normalized to the PAS value obtained in the control tube [43].

Samples were prepared for scanning electron microscopy (SEM) following Merkle et al., with minor modifications [43]. Nonadherent platelets were gently washed away with 1× DPBS, and the samples were immediately fixed with a solution composed of 2% formaldehyde and 2% glutaraldehyde in 1× DPBS for 4 h. The samples were washed with 25% ethanol for 15 min and dehydrated in 50%, 75%, and 100% ethanol at 15-min intervals. The samples were treated with critical point drying in CO_2 using an E3100 Critical Point Dryer (Quorum Technologies LTDA, England). Each sample was then mounted with carbon tape and coated with platinum for 30 s in a Hummer 6.2 argon gas sputter system (Anatech LTD, Battle Creek, MI, USA). All samples were imaged using a field emission SEM (Hitachi, Tokyo, Japan) with the working distance and accelerating voltage set to enhance the contrast of adhered platelets against the cell surfaces. Platelet count on scaffolds was determined using ImageJ. For each manual count, 10 random 40,000 μm^2 sections of area were selected and averaged to estimate the relative adherence of platelets on each scaffold treatment of each cell type [43,44]. Platelets were characterized based on their characteristic size (2.3–4.3 μm) and shape [45]. Aggregated platelets were excluded during platelet counting. Percentage of cell coverage was quantified by segmenting the SEM images and dividing the pixels corresponding to the cell area by the total number of pixels in the image.

2.7. Response to Interleukin 1 Beta (IL-1β) and Endothelial Nitric Oxide Synthase (eNOS) Production

To examine the response to IL-1β, after 4 days of hCB-EC and HUVEC culture in NT or TC scaffolds, recombinant human IL-1β (Life Technologies, Carlsbad, CA, USA) was added to each scaffold at a final concentration of 0.5 ng/mL. After 72-h treatment, the scaffolds with or without IL-1β were fixed with 2% formaldehyde. To examine the eNOS production, a separate set of cultures of hCB-EC and HUVEC in NT or TC scaffolds were fixed with 2% formaldehyde at day 7.

An in-cell enzyme-linked immunosorbent assay (ELISA) was performed using mouse anti-intercellular adhesion molecule 1 (ICAM-1; Abcam, Cambridge, United Kingdom), mouse anti-vascular cell adhesion molecule 1 (VCAM-1; Abcam), and mouse anti-eNOS primary antibodies (Abcam). Primary antibodies were conjugated with HRP rabbit anti-mouse polyclonal secondary antibodies (Abcam) and one-step ultra TMB ELISA (Thermo Fisher Scientific, Waltham, MA, USA) was used to detect HRP activity following the manufacturer's instructions. Absorbance was recorded at 450 nm in a Synergy H1 plate reader (BioTek, Winooski, VT, USA). Background absorbance from the NT and TC scaffolds was obtained from non-seeded scaffolds. The ICAM-1, VCAM-1, and eNOS results were normalized to the average cell number calculated from the MTS calibration curves (Figure S3). Detailed information on the antibodies used is provided in Table S1.

2.8. Statistical Analysis

All values are presented as the mean ± standard deviation unless otherwise specified. The statistical analysis was performed using the software package SPSS (IBM, Armonk, NY, USA). For surface roughness, fiber diameter, scaffold thickness, scaffold porosity, cell proliferation, and ELISA analyses, the statistical analysis was completed using two-way ANOVA with Tukey's range tests to compare between two cell types and two scaffolds. For nitric oxide synthase production, platelet activity assay, and platelet adherence analyses, a one-way ANOVA was performed with a post hoc analysis using individual two-sample two-tailed *t*-tests. The significant difference was determined when the p-value < 0.05.

3. Results

3.1. Cell Characterization

hCB-ECs were derived from the monocytes collected from cord blood. After differentiation, cells were characterized using flow cytometry and immunocytochemistry. Figure 1 shows the results of the flow cytometry represented in histograms. The blue histogram characterizes the signal obtained from the cells treated with the antibodies and the gray histogram represents the cells that were not. A shift in the histogram indicates a relative increase in the average cell fluorescence. Antibodies against CD31 (Figure 1A) and CD105 (Figure 1B) were attached to the cell, confirming the expression of these two endothelial cell-specific markers in our hCB-ECs. In contrast, the blue histogram for the cells treated with the antibodies against CD45 (Figure 1C) did not shift, which can be interpreted as a lack of CD45 expression in our hCB-ECs. These results confirm that hCB-ECs were successfully isolated and there is no contamination from the hematopoietic lineage. Additionally, the immunocytochemistry corroborated the expression of CD31 exclusively in the cell membrane at the cell–cell junctions (Figure 1D).

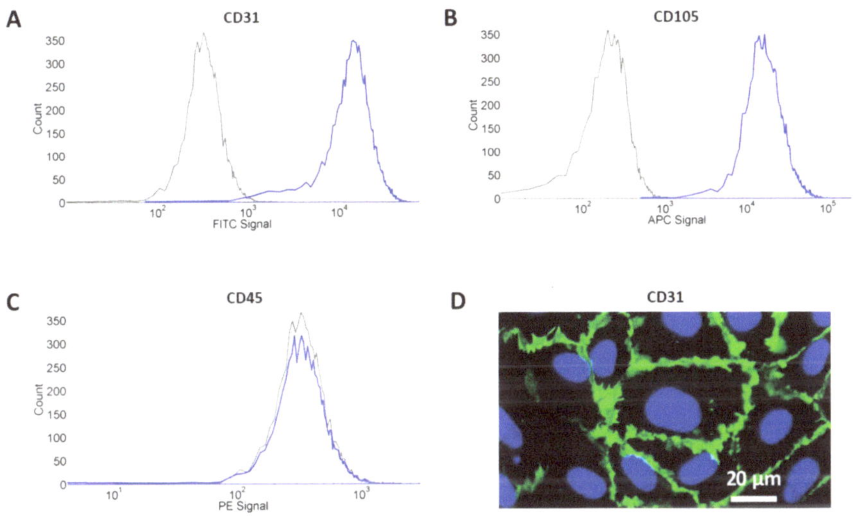

Figure 1. Characterization of human cord blood-derived endothelial cells (hCB-EC) monolayers using flow cytometry and immunocytochemistry. (**A**–**C**) Flow cytometry results show that hCB-ECs are positive for endothelial cell markers CD31 and CD105 while negative for CD45, a hematopoietic marker. (**D**) Immunocytochemistry of CD31 shows that the expression of CD31 is detected on the cell membrane (green—CD31; blue—nuclei). Scale bar = 20 µm.

3.2. Scaffold Characterization

AFM was performed in order to study the effect of thermoforming on surface roughness, which represents the average change in height on the surface with respect to a reference point. In this case, the reference points are the glass coverslips which the individual samples were laid on prior to imaging. No significant difference in roughness was found between the thermoformed and the nontreated scaffolds (299.30 nm ± 27.53 nm vs. 331.53 nm ± 23.44 nm).

Figure 2 summarizes the effect of thermoforming and coating on scaffold microstructure. Figure 2A shows representative SEM images of the nontreated (NT) and thermoformed/coated (TC) scaffolds. The scaffold porosity and fiber diameter were calculated from these SEM images. The histogram of fiber diameter shows a distribution of thinner fibers in the TC scaffolds (Figure 2B), which can be observed in the TC SEM images. A significant reduction in porosity was found in the TC group (36.3% ± 2.39% vs. 41.1% ± 2.54%) (Figure 2C). No significant difference was detected between the thicker fibers of the TC scaffolds and the fibers from NT scaffolds (0.42 µm ± 0.1 µm vs. 0.40 µm ± 0.12 µm). Scaffold thickness was calculated using the Z stacks from the multiphoton images. A significant reduction in thickness was found in the TC group (35.33 µm ± 2.74 µm vs. 61.33 µm ± 7.08 µm) (Figure 2D).

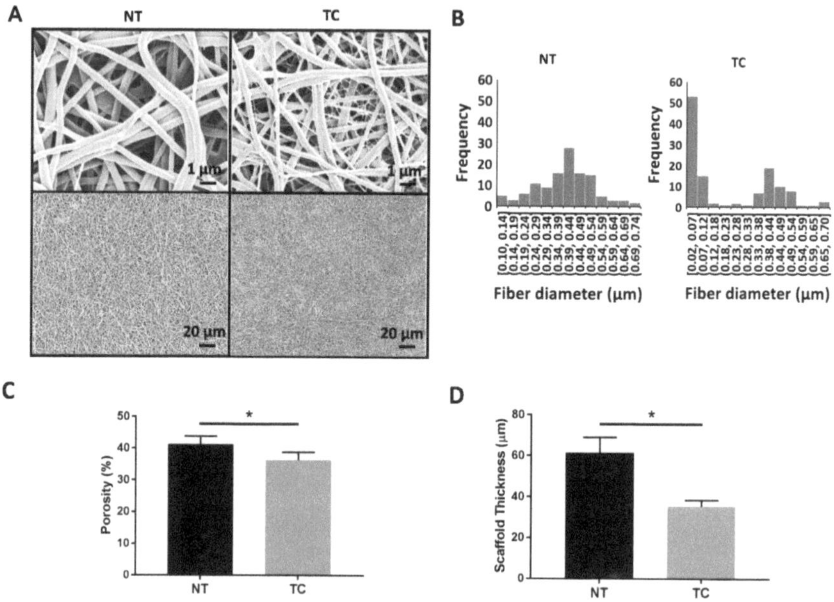

Figure 2. Effect of thermoforming and coating on polycaprolactone/gelatin/fibrinogen scaffolds. (**A**) Representative SEM images of nontreated (NT) or thermoformed and coated (TC) scaffolds. (**B**) Fiber diameter distribution in NT and TC scaffolds was calculated from the SEM images. NT scaffolds have a normal fiber size distribution with fibers ranging from 0.1 to 0.74 μm. The TC scaffolds have a bimodal fiber size distribution with thin fibers ranging from 0.02 to 0.12 μm, and thick fibers ranging from 0.12 to 0.7 μm (n = 6 scaffolds, 120 fibers). (**C**) Average porosity results calculated from the SEM images show a significant decrease in TC scaffolds (* $p < 0.05$; n = 6 scaffolds). (**D**) A significant reduction in thickness calculated from multiphoton images was found in the TC group (* $p < 0.05$; n = 6 scaffolds).

3.3. Effect of Surface Modification on Cell Attachment and Cell Growth

To evaluate the effect of thermoforming and coating on endothelial cell growth, cell number of hCB-ECs and HUVECs in NT or TC scaffolds was evaluated at day 7 using MTS assays (n = 3 in each group). A significant increase in cell number was observed for hCB-ECs compared to HUVECs in NT ($4.9 \times 10^3 \pm 432.04$ vs. $3.04 \times 10^3 \pm 102.6$, $p = 0.003$) and TC scaffolds ($5.71 \times 10^3 \pm 286.7$ vs. $3.40 \times 10^3 \pm 102.06$, $p = 0.001$) (Figure 3).

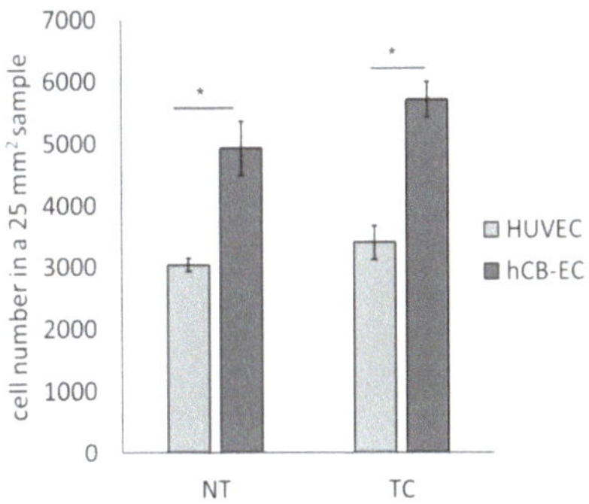

Figure 3. Cell number of hCB-ECs and human umbilical vein endothelial cells (HUVECs) cultured in NT or TC scaffolds at day 7. The cell number was calculated from the MTS calibration curves. A significant increase in hCB-ECs was found in NT and TC scaffolds when compared to HUVECs (* $p < 0.05$, $n = 3$).

To evaluate cell attachment, the cells in scaffolds were stained with DAPI (nuclei) and Phalloidin (F-actin) at day 7. Representative images of maximum intensity projections (MIPs) from each group is presented in Figure 4. We found that the combined surface modification of thermoforming and coating favor cell spreading and cell attachment compared to the NT scaffolds. When we calculated the percentage of cell infiltration from the multiphoton z-stack images, no significant difference was found between hCB-ECs cultured in TC and NT scaffolds (20.40% ± 4.5% vs. 17.2% ± 3.72%) or HUVECs cultured in TC and NT scaffolds (27.01% ± 4.34% vs. 26.01% ± 1.81%). In addition, no significant difference in infiltration was observed between hCB-ECs and HUVECs cultured in TC scaffolds (20.40% ± 4.5% vs. 27.01% ± 4.34%) or NT scaffolds (17.2% ± 3.72% vs. 26.01% ± 1.81%). Interestingly, we noticed that hCB-ECs have a lower cell infiltration than HUVECs in both NT and TC scaffolds.

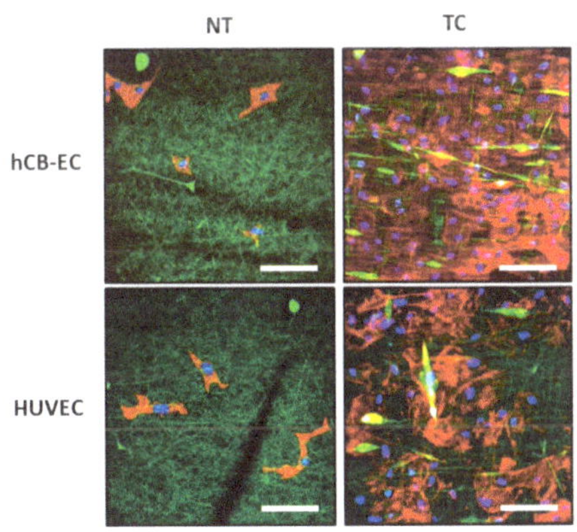

Figure 4. Maximum intensity projection multiphoton images of hCB-ECs or HUVECs cultured in nontreated (NT) or thermoformed/coated (TC) scaffolds after 7 days of culture (green—scaffolds; blue—nuclei; red—F-actin). The combined surface modification of thermoforming and coating favor cell spreading and cell attachment, especially for hCB-ECs. Scale bar = 100 µm.

3.4. *Platelet Activation and Adhesion*

The PAS was calculated for platelets on NT or TC scaffolds seeded with hCB-ECs or HUVECs using a prothrombinase assay. A sample of diluted platelets was pipetted on top of the seeded scaffolds and incubated for 0, 1, and 2 h. After incubation, factor Xa, Ca^{2+}, and acetylated prothrombin were added to the samples and the generation of thrombin as a measurement of platelet activation was quantified by spectrophotometry. PAS values were defined as the ratio of thrombin generated by the samples to thrombin generated by sonication. In Figure 5, we found that in the TC samples, regardless of the cell type, PAS values decrease after 1 h indicating a possible deactivation of platelets. On the contrary, in the NT scaffolds, the PAS values increased over time suggesting a constant activation of platelets. Comparisons were made between scaffolds within the same cell type and between cell types within the same scaffold. A significant difference in absorbance for thrombin generation was found for the NT and TC scaffolds seeded with hCB-ECs at 1 h (0.049 $\pm$ 0.003 vs. 0.026 $\pm$ 0.001, $p = 0.0002$) and 2 h (0.05 $\pm$ 0.003 vs. 0.0049 $\pm$ 0.002, $p = 2.03 \times 10^{-10}$) of platelet–prothrombinase incubation. Also, when comparing the thrombin absorbance results for the NT and TC scaffolds seeded with HUVECs, a significant difference in absorbance was found at 0 h (0.081 $\pm$ 0.01 vs. 0.024 $\pm$ 0.002, $p = 0.001$), 1 h (0.12 $\pm$ 0.04 vs. 0.024 $\pm$ 0.001, $p = 9.83 \times 10^{-7}$), and 2 h (0.135 $\pm$ 0.03 vs. 0.019 $\pm$ 0.001, $p = 4.3 \times 10^{-10}$). A significant decrease in thrombin generation was found in the NT scaffolds seeded with hCB-ECs compared to the ones with HUVECs at 0 h (0.018 $\pm$ 0.001 vs. 0.081 $\pm$ 0.01, $p = 0.001$), 1 h (0.049 $\pm$ 0.003 vs. 0.024 $\pm$ 0.001, $p = 5.46 \times 10^{-7}$), and 2 h (0.05 $\pm$ 0.003 vs. 0.135 $\pm$ 0.002, $p = 2.44 \times 10^{-9}$). For the TC scaffolds seeded with hCB-ECs, the thrombin generation by the perfused platelets was significantly lower when compared to the TC scaffolds seeded with HUVECs after 2 h of incubation (0.0049 $\pm$ 0.002 vs. 0.019 $\pm$ 0.001, $p = 4.3 \times 10^{-10}$).

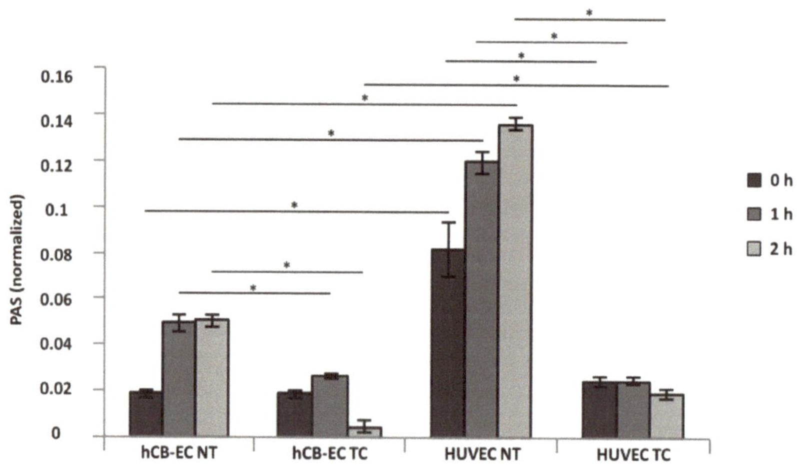

Figure 5. Platelet activity state (PAS) of hCB-EC or HUVEC seeded on thermoformed and coated (TC) scaffolds compared to nontreated (NT) scaffolds at 0, 1, and 2 h. TC groups have lower platelet activation when compared to NT groups; hCB-ECs have lower platelet activation than that of HUVECs (comparisons between hCB-EC and HUVEC cultured in same scaffold and comparisons between NT and TC of same cell type, * $p < 0.05$, $n = 4$).

Platelet adhesion is presented in Figure 6. Figure 6A shows representative SEM images taken for each replicate of NT and TC scaffolds seeded with either hCB-ECs or HUVECs. Cells seeded on NT scaffolds were not able to cover the entire surface, suggesting the presence of material fibers where many platelets are deposited. On the contrary, in TC scaffolds the cells were able to cover almost the entire surface of leaving almost no material fibers exposed reducing the adhesion of platelets. In NT scaffolds, HUVECs have fewer cells attached to the surface than hCB-ECs, leaving more fibers exposed. Similarly, in TC scaffolds HUVECs present more gaps in between the cells resulting in higher platelet adhesion. Figure 6B shows the results of the platelet counts from the SEM images. A significant reduction was found in the number of platelets adhered to both NT (186.9 $\pm$ 56.39 vs. 242.5 $\pm$ 39.22, $p = 0.025$) and TC (55.4 $\pm$ 7.31 vs. 65.77 $\pm$ 5.51, $p = 0.0033$) scaffolds when hCB-ECs were seeded. When assessing the differences between scaffold types, the TC had a significantly lower number of platelets deposited as compared to the nontreated with either hCB-ECs (55.4 ± 7.31 vs. 186.9 ± 56.39, $p = 0.000017$) or HUVECs (65.77 ± 5.51 vs. 242.5 ± 39.22, $p = 8.4 \times 10^{-11}$). Figure 6C presents the quantification of cell coverage calculated from the SEM images. A significant increase in cell coverage was found for hCB-ECs growing in TC compared to NT surfaces (99.73% ± 0.053% vs. 66.72% ± 2.07%, $p = 0.00023$). Similarly, HUVEC presented a significant increase in coverage of TC compared to NT surfaces (99.08% ± 0.13% vs. 40.19% ± 5.83%, $p = 0.0003$). hCB-ECs show significantly more covered area than HUVEC seeded in NT scaffolds (66.72% ± 2.07% vs. 40.19% ± 5.83%, $p = 0.037$).

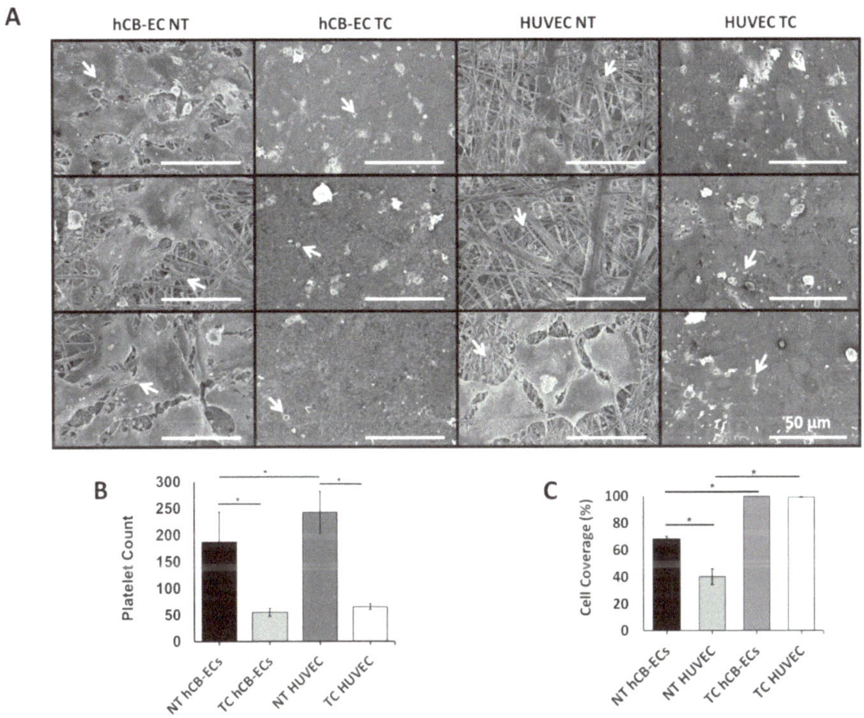

Figure 6. Assessment of platelet adhesion of hCB-ECs or HUVECs cultured on thermoformed/coated (TC) scaffolds compared to nontreated (NT) scaffolds. (**A**) Representative SEM images of each replicate in each experimental group. White arrows are pointing at platelets. Scale bar = 50 µm. (**B**) Platelet counts show that a significant reduction of platelet number in TC groups (* $p < 0.05$, $n = 10$). (**C**) Cell coverage results show that thermoforming and coating increase the cell coverage of scaffolds, and that hCB-ECs have higher cell coverage than HUVECs on NT scaffolds (* $p < 0.05$, $n = 3$).

3.5. Inflammatory Response and eNOS Production

To determine if the cells growing on TC scaffolds can produce eNOS in static conditions, the expression of eNOS was quantified after 7 days in culture using in-cell ELISA. A baseline for eNOS expression was established using NT scaffolds. The absorbance at 450 nm as a result of eNOS production was normalized to the absorbance of MTS assay (490 nm) of hCB-ECs and HUVEC after 7 days growing in the TC and NT scaffolds. The eNOS production was found to be significantly lower in hCB-ECs as compared to HUVECs (6.14 ± 1.04 vs. 11.61 ± 1.04; $p = 0.024$) when the cells are seeded on NT scaffolds (Figure 7A). The production of eNOS by HUVECs growing in the TC scaffolds is significantly lower as compared to their eNOS production when these cells are seeded in NT surfaces (5.52 ± 0.48 vs. 11.61 ± 1.04; $p = 0.006$).

To study the inflammatory response of hCB-ECs and HUVEC on the TC scaffolds, IL-1β was added to the culture and the expression of ICAM-1 and VCAM-1 was quantified using in-cell ELISA. For the hCB-ECs and HUVECs seeded onto TC scaffolds, the addition of IL-1β increased the absorbance at 450 nm. An increase in absorbance is correlated with the expression of inflammation associated proteins VCAM-1 and ICAM-1. All absorbance values were normalized by the MTS assay (490 nm) of hCB-ECs and HUVEC after 7 days growing in the TC scaffolds. A significant decrease in VCAM-1 production was found comparing hCB-ECs to HUVECs after the addition of IL-1β (1.47 ± 0.17 vs. 2.34 ± 0.17; $p = 0.00058$) (Figure 7B). The expression of ICAM-1 is also significantly lower in hCB-EC as compared to HUVEC seeded scaffolds without (0.65 ± 0.11 vs. 1.01 ± 0.089; $p = 0.0015$) and with IL-1β stimulation (0.86 ± 0.07 vs. 1.56 ± 0.19; $p = 0.00067$) (Figure 7C).

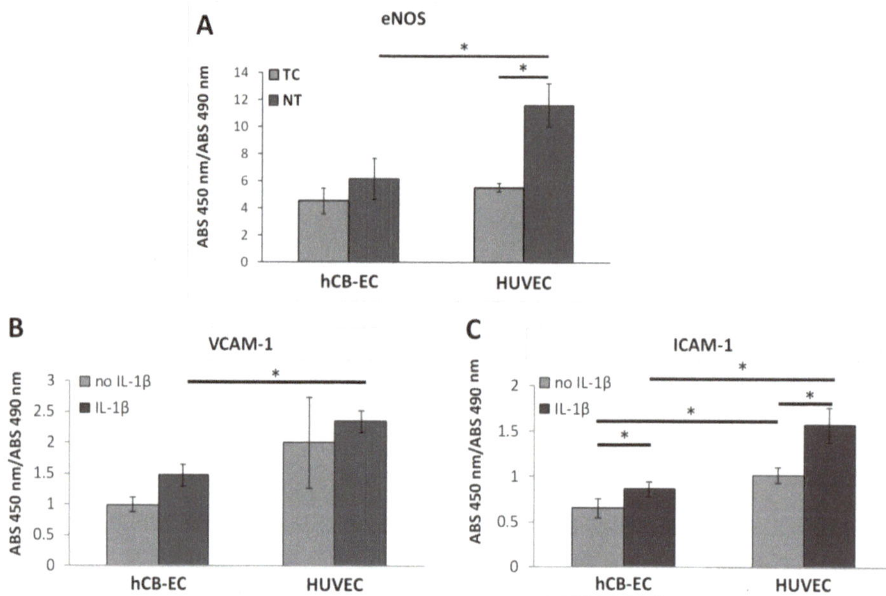

Figure 7. Endothelial nitric oxide synthase (eNOS) production and response to IL-1β of hCB-EC- and HUVEC-seeded scaffolds. (**A**) eNOS ELISA was quantified and normalized to MTS assay. The eNOS production is significantly lower in hCB-ECs as compared to HUVECs when seeded on NT scaffolds (* $p < 0.05$; $n = 4$). (**B**) Anti-vascular cell adhesion molecule 1 (VCAM-1) ELISA in TC scaffolds with or without the addition of 0.5 ng/mL of IL-1β show that HUVECs have higher response compared to hCB-ECs (* $p < 0.05$; $n = 4$). (**C**) ELISA of anti-intercellular adhesion molecule 1 (ICAM-1) shows that HUVECs have higher response compared to hCB-ECs (* $p < 0.05$; $n = 4$).

4. Discussion

The purpose of this study was to evaluate the surface modification of a biomaterial to promote endothelialization of electrospun scaffolds by thermoforming and coating with collagen IV and fibronectin and to quantify the performance of cord blood endothelial cells on our biomaterial surface and its potential for vascular tissue engineering applications.

Our research team was able to isolate mononuclear cells from human umbilical cord blood, differentiate them into endothelial cells, and culture hCB-ECs on electrospun scaffolds while maintaining endothelial cell phenotype. Our findings suggest that scaffolds composed of polycaprolactone/gelatin/fibrinogen that have been surface modified by thermoforming and coating with a mixture 1:1 collagen IV to fibronectin will promote the formation of an endothelial monolayer.

In these scaffolds cell growth is encouraged, resulting in a monolayer of endothelial cells. hCB-ECs proliferate more robustly on our surface modified scaffolds than HUVECs, and the maintenance of the endothelial cell function of both hCB-ECs and HUVECs is comparable. hCB-EC-seeded scaffolds upregulated the adhesion molecules ICAM-1 and VCAM-1 when stimulated with the pro-inflammatory cytokine IL-1β. Our hCB-ECs also displayed the ability to produce the vasoregulatory- and hemocompatibility-related enzyme eNOS and reduce platelet adhesion and activation.

Our source of endothelial cells is cord blood, which is characterized by a unique richness in highly proliferative stem and progenitor cells [46]. Cord blood is readily available, can be collected noninvasively without risk to the mother or infant donor, and can be tested and preserved for long periods of time for future use [47]. This source of endothelial cells has many advantages compared to bone marrow and peripheral blood. Some of the main advantages are the proliferative capacity of the cord blood progenitor cells, higher number of stem cells per volume of blood, and a higher tolerance of cord blood progenitor cells of human leukocyte antigen mismatches [48]. Moreover, hCB-ECs can be used in the treatment of acute ischemic disease, in aging patients, and patients with risk of cardiovascular disease, as opposed to peripheral blood derived endothelial cells [48]. The endothelial cells from cord blood will be treated as a nonautologous source of cells that has to be donor matched [31,49]. Generally, it has been accepted that human leukocyte antigen mismatches are better tolerated with cells derived from cord blood as oppose to bone marrow, resulting in less graft versus host disease [48]. This suggests that endothelial cells derived from cord blood are more adaptable as they are young and not fully differentiated in early passages, therefore they are suitable for clinical applications.

Different efforts have been made for the endothelialization of small diameter vascular grafts, and many of them have used surface modified scaffolds fabricated with synthetic polymers which have excellent biomechanical properties but low biocompatibility [5,50–58]. Surface modifications can reduce hydrophobicity of the material and therefore increase biocompatibility. One of the most popular techniques for improving biocompatibility and endothelial cell adhesion is immobilization of peptide ligands onto the grafts. Peptide sequences such as RGD, GRGDSP, and DGEA have been utilized since they interact directly with endothelial cell receptors and increase cell attachment [59].

In this work, we demonstrated that a scaffold made of a blend of synthetic and natural polymers including polycaprolactone, gelatin, and fibrinogen following by thermoforming and coating with a mixture of collagen IV and fibronectin improves hCB-EC cell growth. This unique hybrid biomaterial possesses the RGD peptide from gelatin [60,61], fibrinogen [62], and fibronectin [63]; GRGDSP peptide from fibronectin [64]; DGEA peptide from gelatin [65]; and FYFDLR from collagen IV [66]. All these are important peptides recognized by integrins in the endothelial cell membrane [66–69]. Furthermore, the inclusion of polycaprolactone to the polymeric blend reduced fiber diameter which appears to promote cell attachment [70] and endothelial cell proliferation [71]. The thermoforming process aimed to smooth the surface of electrospun scaffolds. Surprisingly, thermoforming did not significantly alter the surface roughness in this study. However, the combined surface modification of thermoforming and coating decreased both scaffold porosity and scaffold thickness and altered the distribution of fiber diameters with the appearance of small fibers branching out from the major fibers. We observed better cell attachment in the thermoformed and coated scaffolds demonstrating that the surface-modified scaffolds favor the development of an endothelial cell monolayer. We noticed a difference between the cell number reported for the NT group in Figure 3 and that observed in Figure 4. We speculate this may be due to the staining and intensive washing required prior to multiphoton imaging but not required for an MTS assay, which may remove cells not strongly attached to the nontreated scaffolds. Further experiments should be run to confirm if the decreased strength of attachment of cells placed on NT materials is responsible for the lower cells present in Figure 4. Should this be the case, choosing the TC group would be advantageous given the increased levels of shear stress expected in vivo.

It is possible that the pressure applied to the scaffolds by thermoforming may lead to the production of a more compact fiber arrangement that aids with the retention of coating proteins and that the appearance of small fibers may be due to the coating process creating more contact points for the cells. Thus, the combined surface modification of thermoforming and coating facilitated better hCB-EC attachment and cell growth. Future work will be beneficial to understand the mechanisms by which this process is governed.

Our goal is not only to promote cell attachment and cell growth of hCB-ECs in our electrospun scaffolds but also to assess whether the attached cells can function as mature endothelial monolayers. For this purpose, we evaluated three important characteristics of a healthy endothelial cell layer under static conditions: the response to a pro-inflammatory stimulus, the production of eNOS (enzyme directly related with nitric oxide generation), and the antithrombotic capacity. We investigated the inflammatory response by exposing the hCB-EC- and HUVEC-seeded scaffolds to a pro-inflammatory cytokine IL-1β by assessing the production of VCAM-1 and ICAM-1. Our results show that in both cell types VCAM-1 and ICAM-1 were upregulated, demonstrating a positive response to IL-1β as occurring in the healthy vasculature [72–75]. Interestingly, hCB-ECs intrinsically have a lower production of the VCAM-1 and ICAM-1 which correlates to reduced recruitment of leucocytes/macrophages and thus a reduced probability of graft intimal hyperplasia and atherosclerosis in vivo [76,77].

In endothelial cells, eNOS is responsible for endothelium-derived nitric oxide (NO) production. In this study we measured eNOS production of hCB-ECs or HUVECs seeded on thermoformed/coated scaffolds and compared this with that of cells seeded on nontreated scaffolds. Both cell types expressed eNOS in static conditions in both the thermoformed/coated and nontreated scaffolds indicating that their endothelial cell phenotype is preserved. eNOS production in hCB-ECs is lower than the amount produced by HUVECs in both growing surfaces. These results are similar to those reported by Brown et al. (2009) who showed that hCB-ECs produced a significantly lower amount of eNOS than aortic ECs in static conditions [30]. Similarly, in the work of Yuan et al. (2016), the authors found that endothelial progenitor cells have a markedly lower expression of eNOS as well as activity levels compare to HUVECs and aortic ECs. They concluded that the expression of low levels of eNOS in endothelial progenitor cells compared to mature endothelial cells is due to their higher extracellular matrix deposition [78]. In our work, we also observed that eNOS is reduced significantly for HUVECs growing on TC scaffolds compared to NT scaffolds. This may due to the use of fibronectin in our coating causing the partial downregulation of eNOS in HUVECs which is in line with the study by Viji et al. (2009), who found that HUVECs growing on fibronectin-coated glass slides at a concentration of 50 μm/mL have a significantly reduced eNOS activity when compared to noncoated glass slides [79]. Yuan et al. also found that HUVEC growing in fibronectin-coated flasks have a significant reduction in eNOS mRNA and protein levels compared to HUVEC cultured on polystyrene flasks [78]. In vivo, the predominant physiological stimulus for eNOS active phosphorylation and subsequent NO segregation is wall shear stress [80–82]. It has been demonstrated that eNOS is produced in fewer quantities in endothelial cells cultured in static conditions [80,83,84]. In our work, no significant difference was found between the eNOS protein levels of hCB-ECs growing in NT and TC scaffolds, but it is possible that eNOS production may increase in hCB-EC-seeded scaffolds under flow conditions [30,85].

We also assessed antithrombotic capacity by evaluating platelet activation and deposition when in contact with either nontreated or thermoformed/coated scaffolds preseeded with hCB-ECs or HUVECs. Figure 5 shows the results of the platelet activity state (PAS) assay where the PAS values represent the amount of acetylated thrombin formed by the platelets in contact with the studied samples [43]. Overall, the PAS values over the course of 2 h are lower for the TC scaffolds showing a deactivation after the first hour. It is known that platelets are deactivated by molecules secreted by the endothelium such as NO, prostacyclin, and prostaglandin D [86], and that under normal conditions the coagulation cascade is self-regulated. Furthermore, molecules secreted by platelets and involved in further platelet activation, such as platelet activating factor, thrombin, and tumor necrosis factor-α (TNF-α), trigger the production of NO by the endothelial monolayer resulting in a negative feedback system which deactivates platelets [87–89]. The PAS results correlated with the number of platelets deposited in the scaffolds. There is a significantly lower number of platelets adhered to cell seeded TC scaffolds because of the development of a confluent monolayer. Even though HUVECs formed a monolayer on the TC samples, there were some material fibers exposed for this group likely promoted platelet adhesion. In the nontreated scaffolds, the cells did not cover the entire scaffold, leading to large areas where the material fibers were exposed and the rapid deposition of platelets. According to our findings, there is an inverse relationship between endothelial cell coverage and platelet deposition on the surface. The platelet counts are higher in the scaffolds where the cells are not confluent, and the material is uncovered. Previous studies have shown that endothelial progenitor cells such and hCB-ECs are able to inhibit the attachment and activation of platelets in vitro [90,91]. Moreover, our results are also similar to the study by Brown et al. (2010) who demonstrated that hCB-ECs have superior adhesion and proliferation to vessel-derived cells such as human aortic endothelial cells, and as such, were able to reduce the platelet adhesion and prevent thrombosis in a vein graft [47]. Future studies in our laboratory will utilize immunocytochemistry to further quantify the expression of activated platelet specific markers (e.g., CD62) as well as investigate the effects of shear induced platelet activation using a physiologically realistic pulsatile flow experimental setup.

It should be noted that we did not quantify any of our outcomes on electrospun constructs thermoformed and treated with collagen IV or fibronectin individually. The choice of studying our outcomes on thermoformed and electrospun constructs simultaneously coated with both collagen IV and fibronectin was made after observing the data in Figure S1, which demonstrates that this coating had the highest trend of cell count on coverslips. Therefore, a limitation of our study is that we do not know if the individual coating of collagen IV or fibronectin following thermoforming would have resulted in similar, better, or worse coverage as seen in Figure S2. Another limitation of our study is that all cell experiments were completed under static conditions. Although in vitro endothelialization studies of electrospun scaffolds are primarily performed under static conditions [44,92–94], we recognize the importance of evaluating endothelial cell function under physiological shear stress, an important mechanical stimulus controlling cell response [95]. For example, one of the most important functions of endothelial cells in the vasculature is the production of NO, a potent vasodilator and anti-inflammatory mediator [96]. It is reported that shear stress stimulates eNOS production, eNOS phosphorylation, and NO synthesis [97–101]. Therefore, to be able to make further conclusions about the preservation of endothelial cell phenotype of hCB-EC seeded in thermoformed and coated scaffolds, future studies will focus on elucidating growth and functionality under physiological shear stress. Another limitation of our study is that all scaffolds reported were flat sheets and not tubular scaffolds. One of our ongoing studies is to assess hCB-EC cell growth and function in tubular scaffolds, both in static and flow conditions. Future pulsatile flow experiments will be performed in a parallel plate flow chamber for planar scaffolds and in custom tubular bioreactors.

5. Conclusions

hCB-ECs cultured on scaffolds composed of polycaprolactone, gelatin, and fibrinogen that are thermoformed and coated with collagen IV and fibronectin are attractive for endothelial cell attachment and cell growth. These cells have the capability to proliferate and form a stable monolayer on the surface of the planar scaffold. Upon monolayer formation, hCB-ECs can produce eNOS, respond to the addition of IL-1β through the upregulation of VCAM-1 and ICAM-1, and reduce platelet deposition and activation rate. Our hCB-ECs have superior cell proliferation but lower inflammatory response when compared to HUVECs in electrospun scaffolds.

The long-term goal of our team is to generate a TEVG that supports endothelial cell growth, is biomechanically matched to a target artery, and remains functional and durable post-implantation for artery replacement. With the present work, we demonstrate that thermoforming and coating on polycaprolactone/gelatin/fibrinogen scaffolds enhances hCB-EC growth and improves maintenance of their EC phenotype in vitro. Future directions will include the development and assessment of layered scaffolds as well as in vivo studies using an aortic implantation to functionally assess our TEVG.

We believe this work provides important findings towards the endothelialization of electrospun scaffolds, and thus will have an impact on the field of tissue engineering and regenerative medicine. We demonstrated that hybrid electrospun scaffolds that were surface modified by the combined treatments of thermoforming and coating and were subsequently seeded with endothelial cells derived from cord blood show promise for vascular tissue engineering.

Supplementary Materials: The following are available online at http://www.mdpi.com/2077-0383/8/2/185/s1, Table S1. Detailed information on antibodies, Figure S1. Cell number of hCB-ECs 3 days post seeding on glass micro-coverslips coated with collagen type I (ColI), collagen type IV (ColIV), fibronectin (Fib), or a 1:1 mixture of ColI and ColIV (ColI/ColIV), ColI and Fib (ColI/Fib), or ColIV and Fib (ColIV/Fib) in 96-well plates, Figure S2. Maximum intensity projection multiphoton images of hCB-ECs or HUVECs cultured in nontreated, thermoformed, coated, or thermoformed coated scaffolds with collagen IV and fibronectin after 7 days of culture (green—scaffolds; blue—nuclei; red—F-actin), Figure S3. MTS assay calibration curves for hCB-ECs and HUVEC ($n = 6$).

Author Contributions: Study Conception and Design: D.C.A. and J.P.V.G.; Acquisition of Data: D.C.A., J.-J.L., and D.M.; Analysis and Interpretation of Data: D.C.A., J.-J.L., and D.M.; Drafting of Manuscript: D.C.A.; Critical Revision: D.C.A., J.-J.L., D.M., M.J.S., M.B., W.R.W., D.H., and J.P.V.G.

Funding: This research was funded by the NIH (1R56HL136517-01, PI: J.P.V.G.). D.C.A. was also funded by the Cellular Approaches to Tissue Engineering and Regeneration training grant (CATER) (5T32EB001026) at the University of Pittsburgh. All Cypher AFM images and data were collected on the W.M. Keck Center for Surface and Interface Imaging at the University of Arizona supported by NSF (1337371).

Conflicts of Interest: The authors declare no conflicts of interest. The funders had no role in the design of the study; in the collection, analyses, or interpretation of data; in the writing of the manuscript, and in the decision to publish the results.

References

1. Roger, V.L.; Go, A.S.; Lloyd-Jones, D.M.; Benjamin, E.J.; Berry, J.D.; Borden, W.B.; Bravata, D.M.; Dai, S.; Ford, E.S.; Fox, C.S. Heart disease and stroke statistics—2012 update a report from the American heart association. *Circulation* **2012**, *125*, e2–e220. [PubMed]
2. Mozaffarian, D.; Benjamin, E.J.; Go, A.S.; Arnett, D.K.; Blaha, M.J.; Cushman, M.; Das, S.R.; de Ferranti, S.; Després, J.-P.; Fullerton, H.J. Executive summary: Heart disease and stroke statistics—2016 update a report from the American heart association. *Circulation* **2016**, *133*, 447–454. [CrossRef] [PubMed]
3. Sell, S.A.; McClure, M.J.; Garg, K.; Wolfe, P.S.; Bowlin, G.L. Electrospinning of collagen/biopolymers for regenerative medicine and cardiovascular tissue engineering. *Adv. Drug Deliv. Rev.* **2009**, *61*, 1007–1019. [CrossRef] [PubMed]
4. Go, A.S.; Mozaffarian, D.; Roger, V.L.; Benjamin, E.J.; Berry, J.D.; Blaha, M.J.; Dai, S.; Ford, E.S.; Fox, C.S.; Franco, S. Heart disease and stroke statistics—2014 update: A report from the American Heart Association. *Circulation* **2014**, *129*, e28–e292. [CrossRef] [PubMed]
5. McClure, M.J.; Sell, S.A.; Simpson, D.G.; Walpoth, B.H.; Bowlin, G.L. A three-layered electrospun matrix to mimic native arterial architecture using polycaprolactone, elastin, and collagen: A preliminary study. *Acta Biomater.* **2010**, *6*, 2422–2433. [CrossRef] [PubMed]
6. McClure, M.; Wolfe, P.; Rodriguez, I.; Bowlin, G. Bioengineered vascular grafts: Improving vascular tissue engineering through scaffold design. *J. Drug Deliv. Sci. Technol.* **2011**, *21*, 211–227. [CrossRef]
7. Byrom, M.J.; Ng, M.K.; Bannon, P.G. Biomechanics and biocompatibility of the perfect conduit—Can we build one? *Ann. Cardiothorac. Surg.* **2013**, *2*, 435–443.
8. L'Heureux, N.; Dusserre, N.; Marini, A.; Garrido, S.; de la Fuente, L.; McAllister, T. Technology insight: The evolution of tissue-engineered vascular grafts—From research to clinical practice. *Nat. Clin. Pract. Cardiovasc. Med.* **2007**, *4*, 389–395. [CrossRef]
9. Tan, A.; Gundogan, B.; Farhatnia, Y.; Nayyer, L.; Mahdibeiraghdar, S.; Rajadas, J.; De Coppi, P.; Davies, A.H.; Seifalian, A.M. Tissue engineering vascular grafts a fortiori: Looking back and going forward. *Expert Opin. Biol. Ther.* **2015**, *15*, 231–244.
10. Hasan, A.; Memic, A.; Annabi, N.; Hossain, M.; Paul, A.; Dokmeci, M.R.; Dehghani, F.; Khademhosseini, A. Electrospun scaffolds for tissue engineering of vascular grafts. *Acta Biomater.* **2014**, *10*, 11–25. [CrossRef]

11. Ravi, S.; Chaikof, E.L. Biomaterials for vascular tissue engineering. *Regen. Med.* **2010**, *5*, 107–120. [CrossRef] [PubMed]
12. Dhandayuthapani, B.; Yoshida, Y.; Maekawa, T.; Kumar, D.S. Polymeric scaffolds in tissue engineering application: A review. *Int. J. Polym. Sci.* **2011**, *2011*. [CrossRef]
13. Lannutti, J.; Reneker, D.; Ma, T.; Tomasko, D.; Farson, D. Electrospinning for tissue engineering scaffolds. *Mater. Sci. Eng. C* **2007**, *27*, 504–509. [CrossRef]
14. Rim, N.G.; Shin, C.S.; Shin, H. Current approaches to electrospun nanofibers for tissue engineering. *Biomed. Mater.* **2013**, *8*, 014102. [CrossRef] [PubMed]
15. Nagiah, N.; Johnson, R.; Anderson, R.; Elliott, W.; Tan, W. Highly compliant vascular grafts with gelatin-sheathed coaxially structured nanofibers. *Langmuir* **2015**, *31*, 12993–13002. [CrossRef]
16. Fukunishi, T.; Best, C.A.; Sugiura, T.; Shoji, T.; Yi, T.; Udelsman, B.; Ohst, D.; Ong, C.S.; Zhang, H.; Shinoka, T. Tissue-engineered small diameter arterial vascular grafts from cell-free nanofiber pcl/chitosan scaffolds in a sheep model. *PLoS ONE* **2016**, *11*, e0158555. [CrossRef] [PubMed]
17. Pan, Y.; Zhou, X.; Wei, Y.; Zhang, Q.; Wang, T.; Zhu, M.; Li, W.; Huang, R.; Liu, R.; Chen, J. Small-diameter hybrid vascular grafts composed of polycaprolactone and polydioxanone fibers. *Sci. Rep.* **2017**, *7*, 3615. [CrossRef] [PubMed]
18. Tamimi, E.; Ardila, D.; Haskett, D.; Doetschman, T.; Slepian, M.J.; Kellar, R.S.; Geest, J.V. Biomechanical comparison of glutaraldehyde-crosslinked gelatin fibrinogen electrospun scaffolds to porcine coronary arteries. *J. Biomech. Eng.* **2016**, *138*, 011001. [CrossRef] [PubMed]
19. Melchiorri, A.; Hibino, N.; Yi, T.; Lee, Y.; Sugiura, T.; Tara, S.; Shinoka, T.; Breuer, C.; Fisher, J. Contrasting biofunctionalization strategies for the enhanced endothelialization of biodegradable vascular grafts. *Biomacromolecules* **2015**, *16*, 437–446. [CrossRef]
20. Waller, B.F.; Orr, C.M.; Slack, J.D.; Pinkerton, C.A.; Van Tassel, J.; Peters, T. Anatomy, histology, and pathology of coronary arteries: A review relevant to new interventional and imaging techniques—Part I. *Clin. Cardiol.* **1992**, *15*, 451–457. [CrossRef]
21. Michiels, C. Endothelial cell functions. *J. Cell. Physiol.* **2003**, *196*, 430–443. [CrossRef] [PubMed]
22. Deanfield, J.E.; Halcox, J.P.; Rabelink, T.J. Endothelial function and dysfunction testing and clinical relevance. *Circulation* **2007**, *115*, 1285–1295. [CrossRef] [PubMed]
23. Zhang, H.; Tao, Y.; Ren, S.; Liu, H.; Zhou, H.; Hu, J.; Tang, Y.; Zhang, B.; Chen, H. Isolation and characterization of human umbilical cord-derived endothelial colony-forming cells. *Exp. Ther. Med.* **2017**, *14*, 4160–4166. [CrossRef] [PubMed]
24. Weber, B.; Zeisberger, S.M.; Hoerstrup, S.P. Umbilical cord blood-derived endothelial progenitor cells for cardiovascular tissue engineering. In *Perinatal Stem Cells*, 1st ed.; Springer: New York, NY, USA, 2014; pp. 325–336.
25. Stroncek, J.D.; Grant, B.S.; Brown, M.A.; Povsic, T.J.; Truskey, G.A.; Reichert, W.M. Comparison of endothelial cell phenotypic markers of late-outgrowth endothelial progenitor cells isolated from patients with coronary artery disease and healthy volunteers. *Tissue Eng. Part A* **2009**, *15*, 3473–3486. [CrossRef] [PubMed]
26. Ingram, D.A.; Mead, L.E.; Tanaka, H.; Meade, V.; Fenoglio, A.; Mortell, K.; Pollok, K.; Ferkowicz, M.J.; Gilley, D.; Yoder, M.C. Identification of a novel hierarchy of endothelial progenitor cells using human peripheral and umbilical cord blood. *Blood* **2004**, *104*, 2752–2760. [CrossRef]
27. Cai, H.; Gehrig, P.; Scott, T.M.; Zimmermann, R.; Schlapbach, R.; Zisch, A.H. MnSOD marks cord blood late outgrowth endothelial cells and accompanies robust resistance to oxidative stress. *Biochem. Biophys. Res. Commun.* **2006**, *350*, 364–369. [CrossRef] [PubMed]
28. Bompais, H.; Chagraoui, J.; Canron, X.; Crisan, M.; Liu, X.H.; Anjo, A.; Tolla-Le Port, C.; Leboeuf, M.; Charbord, P.; Bikfalvi, A. Human endothelial cells derived from circulating progenitors display specific functional properties compared with mature vessel wall endothelial cells. *Blood* **2004**, *103*, 2577–2584. [CrossRef]
29. Schmidt, D.; Breymann, C.; Weber, A.; Guenter, C.I.; Neuenschwander, S.; Zund, G.; Turina, M.; Hoerstrup, S.P. Umbilical cord blood derived endothelial progenitor cells for tissue engineering of vascular grafts. *Ann. Thorac. Surg.* **2004**, *78*, 2094–2098. [CrossRef]
30. Brown, M.A.; Wallace, C.S.; Angelos, M.; Truskey, G.A. Characterization of umbilical cord blood-derived late outgrowth endothelial progenitor cells exposed to laminar shear stress. *Tissue Eng. Part A* **2009**, *15*, 3575–3587. [CrossRef]

31. Brown, M.E.; Rondon, E.; Rajesh, D.; Mack, A.; Lewis, R.; Feng, X.; Zitur, L.J.; Learish, R.D.; Nuwaysir, E.F. Derivation of induced pluripotent stem cells from human peripheral blood T lymphocytes. *PLoS ONE* **2010**, *5*, e11373. [CrossRef]
32. Jung, Y.; Ji, H.; Chen, Z.; Fai Chan, H.; Atchison, L.; Klitzman, B.; Truskey, G.; Leong, K.W. Scaffold-free, human mesenchymal stem cell-based tissue engineered blood vessels. *Sci. Rep.* **2015**, *5*, 15116. [CrossRef]
33. Javed, M.J.; Mead, L.E.; Prater, D.; Bessler, W.K.; Foster, D.; Case, J.; Goebel, W.S.; Yoder, M.C.; Haneline, L.S.; Ingram, D.A. Endothelial colony forming cells and mesenchymal stem cells are enriched at different gestational ages in human umbilical cord blood. *Pediatr. Res.* **2008**, *64*, 68–73. [CrossRef]
34. Balasubramanian, P.; Prabhakaran, M.P.; Kai, D.; Ramakrishna, S. Human cardiomyocyte interaction with electrospun fibrinogen/gelatin nanofibers for myocardial regeneration. *J. Biomater. Sci. Polym. Ed.* **2013**, *24*, 1660–1675. [CrossRef] [PubMed]
35. Ardila, D.C.; Tamimi, E.; Danford, F.L.; Haskett, D.G.; Kellar, R.S.; Doetschman, T.; Vande Geest, J.P. TGFβ2 differentially modulates smooth muscle cell proliferation and migration in electrospun gelatin-fibrinogen constructs. *Biomaterials* **2015**, *37*, 164–173. [CrossRef] [PubMed]
36. Koynova, R.; Antonova, B.; Sezanova, B.; Tenchov, B.J.T.A. Beneficial effect of sequential chemotherapy treatments of lung cancer patients revealed by calorimetric monitoring of blood plasma proteome denaturation. *Thermochim. Acta* **2018**, *659*, 1–7. [CrossRef]
37. Biscarat, J.; Charmette, C.; Sanchez, J.; Pochat-Bohatier, C. Preparation of dense gelatin membranes by combining temperature induced gelation and dry-casting. *J. Membr. Sci.* **2015**, *473*, 45–53. [CrossRef]
38. Ashton, J.H.; Mertz, J.A.; Harper, J.L.; Slepian, M.J.; Mills, J.L.; McGrath, D.V.; Vande Geest, J.P. Polymeric endoaortic paving: Mechanical, thermoforming, and degradation properties of polycaprolactone/polyurethane blends for cardiovascular applications. *Acta Biomater.* **2011**, *7*, 287–294. [CrossRef] [PubMed]
39. Sasidharan, A.; David, A.; Gohil, A.; Gupta, A.K. Simple device to determine the pressure applied by pressure clips for the treatment of earlobe keloids. *Indian J. Plast. Surg.* **2015**, *48*, 293–296. [PubMed]
40. Williams, M.J.; Utzinger, U.; Barkmeier-Kraemer, J.M.; Vande Geest, J.P. Differences in the microstructure and biomechanical properties of the recurrent laryngeal nerve as a function of age and location. *J. Biomech. Eng.* **2014**, *136*, 081008. [CrossRef] [PubMed]
41. Haskett, D.; Azhar, M.; Utzinger, U.; Vande Geest, J.P. Progressive alterations in microstructural organization and biomechanical response in the ApoE mouse model of aneurysm. *Biomatter* **2013**, *3*, e24648. [CrossRef] [PubMed]
42. Jesty, J.; Bluestein, D. Acetylated prothrombin as a substrate in the measurement of the procoagulant activity of platelets: Elimination of the feedback activation of platelets by thrombin. *Anal. Biochem.* **1999**, *272*, 64–70. [CrossRef]
43. Merkle, V.M.; Martin, D.; Hutchinson, M.; Tran, P.L.; Behrens, A.; Hossainy, S.; Sheriff, J.; Bluestein, D.; Wu, X.; Slepian, M.J. Hemocompatibility of poly (vinyl alcohol)–gelatin core–shell electrospun nanofibers: A scaffold for modulating platelet deposition and activation. *ACS Appl. Mater. Interfaces* **2015**, *7*, 8302–8312. [CrossRef]
44. Merkle, V.M.; Tran, P.L.; Hutchinson, M.; Ammann, K.R.; DeCook, K.; Wu, X.; Slepian, M.J. Core–shell PVA/gelatin electrospun nanofibers promote human umbilical vein endothelial cell and smooth muscle cell proliferation and migration. *Acta Biomater.* **2015**, *27*, 77–87. [CrossRef] [PubMed]
45. Kuwahara, M.; Sugimoto, M.; Tsuji, S.; Matsui, H.; Mizuno, T.; Miyata, S.; Yoshioka, A. Platelet shape changes and adhesion under high shear flow. *Arterioscler. Thromb. Vasc. Biol.* **2002**, *22*, 329–334. [CrossRef] [PubMed]
46. Stavropoulos-Giokas, C.; Charron, D.; Navarrete, C. *Cord Blood Stem Cells Medicine*, 1st ed.; Elsevier: Cambridge, MA, USA, 2014.
47. Brown, M.A.; Zhang, L.; Levering, V.W.; Wu, J.-H.; Satterwhite, L.L.; Brian, L.; Freedman, N.J.; Truskey, G.A. Human umbilical cord blood–derived endothelial cells reendothelialize vein grafts and prevent thrombosis. *Arterioscler. Thromb. Vasc. Biol.* **2010**, *30*, 2150–2155. [CrossRef] [PubMed]
48. Zhang, L.; Yang, R.; Han, Z.C. Transplantation of umbilical cord blood-derived endothelial progenitor cells: A promising method of therapeutic revascularisation. *Eur. J. Haematol.* **2006**, *76*, 1–8. [CrossRef] [PubMed]
49. El-Badri, N. Advances in umbilical cord blood therapy: Hematopoietic stem cell transplantation and beyond. In *Advances in Stem Cell Therapy*; El-Badri, N. Springer: New York, NY, USA, 2017; pp. 139–168.

50. Burrows, M.C.; Zamarion, V.M.; Filippin-Monteiro, F.B.; Schuck, D.C.; Toma, H.E.; Campa, A.; Garcia, C.R.; Catalani, L.H. Hybrid scaffolds built from pet and collagen as a model for vascular graft architecture. *Macromol. Biosci.* **2012**, *12*, 1660–1670. [CrossRef]
51. Ma, Z.; Kotaki, M.; Yong, T.; He, W.; Ramakrishna, S. Surface engineering of electrospun polyethylene terephthalate (PET) nanofibers towards development of a new material for blood vessel engineering. *Biomaterials* **2005**, *26*, 2527–2536. [CrossRef]
52. Cassady, A.I.; Hidzir, N.M.; Grøndahl, L. Enhancing expanded poly (tetrafluoroethylene) (ePTFE) for biomaterials applications. *J. Appl. Polym. Sci.* **2014**, *131*, 40533. [CrossRef]
53. Takagi, H.; Goto, S.N.; Matsui, M.; Manabe, H.; Umemoto, T. A contemporary meta-analysis of Dacron versus polytetrafluoroethylene grafts for femoropopliteal bypass grafting. *J. Vasc. Surg.* **2010**, *52*, 232–236. [CrossRef]
54. Jeschke, M.G.; Hermanutz, V.; Wolf, S.E.; Köveker, G.B. Polyurethane vascular prostheses decreases neointimal formation compared with expanded polytetrafluoroethylene. *J. Vasc. Surg.* **1999**, *29*, 168–176. [CrossRef]
55. MüLLER-HüLSBECK, S.; Walluscheck, K.P.; Priebe, M.; Grimm, J.; Cremer, J.; Heller, M. Experience on endothelial cell adhesion on vascular stents and stent-grafts: First in vitro results. *Invest. Radiol.* **2002**, *37*, 314–320. [CrossRef] [PubMed]
56. Ren, X.; Feng, Y.; Guo, J.; Wang, H.; Li, Q.; Yang, J.; Hao, X.; Lv, J.; Ma, N.; Li, W. Surface modification and endothelialization of biomaterials as potential scaffolds for vascular tissue engineering applications. *Chem. Soc. Rev.* **2015**, *44*, 5680–5742. [CrossRef] [PubMed]
57. Li, Q.; Wang, Z.; Zhang, S.; Zheng, W.; Zhao, Q.; Zhang, J.; Wang, L.; Wang, S.; Kong, D. Functionalization of the surface of electrospun poly (epsilon-caprolactone) mats using zwitterionic poly (carboxybetaine methacrylate) and cell-specific peptide for endothelial progenitor cells capture. *Mater. Sci. Eng. C* **2013**, *33*, 1646–1653. [CrossRef] [PubMed]
58. Xiong, G.M.; Yuan, S.; Tan, C.K.; Wang, J.K.; Liu, Y.; Tan, T.T.Y.; Tan, N.S.; Choong, C. Endothelial cell thrombogenicity is reduced by ATRP-mediated grafting of gelatin onto PCL surfaces. *J. Mater. Chem. B* **2014**, *2*, 485–493. [CrossRef]
59. Choi, W.S.; Joung, Y.K.; Lee, Y.; Bae, J.W.; Park, H.K.; Park, Y.H.; Park, J.-C.; Park, K.D. Enhanced patency and endothelialization of small-caliber vascular grafts fabricated by coimmobilization of heparin and cell-adhesive peptides. *ACS Appl. Mater. Interfaces* **2016**, *8*, 4336–4346. [CrossRef] [PubMed]
60. Huang, Y.; Onyeri, S.; Siewe, M.; Moshfeghian, A.; Madihally, S.V. In vitro characterization of chitosan–gelatin scaffolds for tissue engineering. *Biomaterials* **2005**, *26*, 7616–7627. [CrossRef] [PubMed]
61. Wu, S.-C.; Chang, W.-H.; Dong, G.-C.; Chen, K.-Y.; Chen, Y.-S.; Yao, C.-H. Cell adhesion and proliferation enhancement by gelatin nanofiber scaffolds. *J. Bioact. Compatible Polym.* **2011**, *26*, 565–577. [CrossRef]
62. D'Souza, S.E.; Ginsberg, M.H.; Plow, E.F. Arginyl-glycyl-aspartic acid (RGD): A cell adhesion motif. *Trends Biochem. Sci.* **1991**, *16*, 246–250. [CrossRef]
63. Ruoslahti, E.; Pierschbacher, M.D. Arg-gly-asp: A versatile cell recognition signal. *Cell* **1986**, *44*, 517–518. [CrossRef]
64. Widhe, M.; Shalaly, N.D. A fibronectin mimetic motif improves integrin mediated cell biding to recombinant spider silk matrices. *Biomaterials* **2016**, *74*, 256–266. [CrossRef] [PubMed]
65. Pierce, B.F.; Pittermann, E.; Ma, N.; Gebauer, T.; Neffe, A.T.; Hölscher, M.; Jung, F.; Lendlein, A. Viability of human mesenchymal stem cells seeded on crosslinked entropy-elastic gelatin-based hydrogels. *Macromol. Biosci.* **2012**, *12*, 312–321. [CrossRef] [PubMed]
66. Underwood, P.A.; Bennett, F.A.; Kirkpatrick, A.; Bean, P.A.; Moss, B.A. Evidence for the location of a binding sequence for the $\alpha 2\beta 1$ integrin of endothelial cells, in the $\beta 1$ subunit of laminin. *Biochem. J.* **1995**, *309*, 765–771. [CrossRef] [PubMed]
67. Saotome, T.; Hayashi, H.; Tanaka, R.; Kinugasa, A.; Uesugi, S.; Tatematsu, K.-I.; Sezutsu, H.; Kuwabara, N.; Asakura, T. Introduction of VEGF or RGD sequences improves revascularization properties of bombyx mori silk fibroin produced by transgenic silkworm. *J. Mater. Chem. B* **2015**, *3*, 7109–7116. [CrossRef]
68. Ingber, D.E. Fibronectin controls capillary endothelial cell growth by modulating cell shape. *Proc. Natl. Acad. Sci. USA* **1990**, *87*, 3579–3583. [CrossRef] [PubMed]

69. Bahou, W.F.; Potter, C.L.; Mirza, H. The VLA-2 (alpha 2 beta 1) I domain functions as a ligand-specific recognition sequence for endothelial cell attachment and spreading: Molecular and functional characterization. *Blood* **1994**, *84*, 3734–3741. [PubMed]
70. Chen, M.; Patra, P.K.; Warner, S.B.; Bhowmick, S. Role of fiber diameter in adhesion and proliferation of NIH 3T3 fibroblast on electrospun polycaprolactone scaffolds. *Tissue Eng.* **2007**, *13*, 579–587. [CrossRef] [PubMed]
71. Whited, B.M.; Rylander, M.N. The influence of electrospun scaffold topography on endothelial cell morphology, alignment, and adhesion in response to fluid flow. *Biotechnol. Bioeng.* **2014**, *111*, 184–195. [CrossRef] [PubMed]
72. Adams, W.J.; Zhang, Y.; Cloutier, J.; Kuchimanchi, P.; Newton, G.; Sehrawat, S.; Aird, W.C.; Mayadas, T.N.; Luscinskas, F.W.; García-Cardeña, G. Functional vascular endothelium derived from human induced pluripotent stem cells. *Stem Cell Rep.* **2013**, *1*, 105–113. [CrossRef] [PubMed]
73. Van Rijssel, J.; Timmerman, I.; Van Alphen, F.P.; Hoogenboezem, M.; Korchynskyi, O.; Geerts, D.; Geissler, J.; Reedquist, K.A.; Niessen, H.W.; Van Buul, J.D. The Rho-GEF Trio regulates a novel pro-inflammatory pathway through the transcription factor Ets2. *Biol. Open* **2013**, *2*, 569–579. [CrossRef] [PubMed]
74. Takahashi, M.; Ikeda, U.; Masuyama, J.-I.; Kitagawa, S.-I.; Kasahara, T.; Shimpo, M.; Kano, S.; Shimada, K. Monocyte-endothelial cell interaction induces expression of adhesion molecules on human umbilical cord endothelial cells. *Cardiovasc. Res.* **1996**, *32*, 422–429. [CrossRef]
75. Blake, G.J.; Ridker, P.M. Novel clinical markers of vascular wall inflammation. *Circul. Res.* **2001**, *89*, 763–771. [CrossRef]
76. Qin, L.; Huang, Q.; Zhang, H.; Liu, R.; Tellides, G.; Min, W.; Yu, L. Socs1 prevents graft arteriosclerosis by preserving endothelial cell function. *J. Am. Coll. Cardiol.* **2014**, *63*, 21–29. [CrossRef] [PubMed]
77. Zou, Y.; Hu, Y.; Mayr, M.; Dietrich, H.; Wick, G.; Xu, Q. Reduced neointima hyperplasia of vein bypass grafts in intercellular adhesion molecule-1–deficient mice. *Circul. Res.* **2000**, *86*, 434–440. [CrossRef]
78. Yuan, Y.; Stewart, D.J.; Courtman, D.W. The regulation of endothelial nitric oxide synthase by extracellular matrix in human late outgrowth endothelial progenitor cells. In *Front. Bioeng. Biotechnol. Conference Abstract: 10th World Biomaterials Congress, Montréal, Canada, 17–22 May, 2016*; Frontiers in Bioengineering and Biotechnology: Lausanne, Switzerland, 2016.
79. Viji, R.; Kumar, V.S.; Kiran, M.; Sudhakaran, P. Modulation of endothelial nitric oxide synthase by fibronectin. *Mol. Cell. Biochem.* **2009**, *323*, 91–100. [CrossRef] [PubMed]
80. Matsushita, H.; Chang, E.; Glassford, A.J.; Cooke, J.P.; Chiu, C.-P.; Tsao, P.S. eNOS activity is reduced in senescent human endothelial cells preservation by hTERT immortalization. *Circul. Res.* **2001**, *89*, 793–798. [CrossRef]
81. Tran, J.; Magenau, A.; Rodriguez, M.; Rentero, C.; Royo, T.; Enrich, C.; Thomas, S.R.; Grewal, T.; Gaus, K. Activation of endothelial nitric oxide (eNOS) occurs through different membrane domains in endothelial cells. *PLoS ONE* **2016**, *11*, e0151556. [CrossRef]
82. Ruan, T.; Bharath, L.; Mueller, R.; Goodrich, R.; Graham, T.; Symons, J.D. Shear-induced extracellular regulated kinase signaling to eNOS is increased when autophagy is compromised in endothelial cells. *FASEB J.* **2015**, *29*, 956–954.
83. Li, Y.; Zheng, J.; Bird, I.M.; Magness, R.R. Effects of pulsatile shear stress on signaling mechanisms controlling nitric oxide production, endothelial nitric oxide synthase phosphorylation, and expression in ovine fetoplacental artery endothelial cells. *Endothelium* **2005**, *12*, 21–39. [CrossRef]
84. Yang, B.; Rizzo, V. Shear stress activates eNOS at the endothelial apical surface through β1 containing integrins and caveolae. *Cell. Mol. Bioeng.* **2013**, *6*, 346–354. [CrossRef]
85. Do Kang, S.; Carlon, T.A.; Jantzen, A.E.; Lin, F.-H.; Ley, M.M.; Allen, J.D.; Stabler, T.V.; Haley, N.R.; Truskey, G.A.; Achneck, H.E. Isolation of functional human endothelial cells from small volumes of umbilical cord blood. *Ann. Biomed. Eng.* **2013**, *41*, 2181–2192. [CrossRef] [PubMed]
86. Broos, K.; Feys, H.B.; De Meyer, S.F.; Vanhoorelbeke, K.; Deckmyn, H. Platelets at work in primary hemostasis. *Blood Rev.* **2011**, *25*, 155–167. [CrossRef] [PubMed]
87. Szabo, C. Alterations in nitric oxide production in various forms of circulatory shock. *New Horiz.* **1995**, *3*, 2–32. [PubMed]
88. Van Hinsbergh, V.W. Endothelium—Role in regulation of coagulation and inflammation. *Semin. Immunopathol.* **2012**, *34*, 93–106. [CrossRef] [PubMed]

89. Cronenwett, J.L.; Johnston, K.W. *Rutherford's Vascular Surgery*; Elsevier Health Sciences: Amsterdam, The Netherlands, 2014.
90. Abou-Saleh, H.; Yacoub, D.; Théorêt, J.-F.; Gillis, M.-A.; Neagoe, P.-E.; Labarthe, B.; Théroux, P.; Sirois, M.G.; Tabrizian, M.; Thorin, E. Endothelial progenitor cells bind and inhibit platelet function and thrombus formation. *Circulation* **2009**, *120*, 2230–2239. [CrossRef]
91. Shirota, T.; He, H.; Yasui, H.; Matsuda, T. Human endothelial progenitor cell-seeded hybrid graft: Proliferative and antithrombogenic potentials in vitro and fabrication processing. *Tissue Eng.* **2003**, *9*, 127–136. [CrossRef]
92. He, W.; Yong, T.; Teo, W.E.; Ma, Z.; Ramakrishna, S. Fabrication and endothelialization of collagen-blended biodegradable polymer nanofibers: Potential vascular graft for blood vessel tissue engineering. *Tissue Eng.* **2005**, *11*, 1574–1588. [CrossRef]
93. Lee, B.; Shafiq, M.; Jung, Y.; Park, J.-C.; Kim, S.H. Characterization and preparation of bio-tubular scaffolds for fabricating artificial vascular grafts by combining electrospinning and a co-culture system. *Macromol. Res.* **2016**, *24*, 131–142. [CrossRef]
94. Zhou, W.; Feng, Y.; Yang, J.; Fan, J.; Lv, J.; Zhang, L.; Guo, J.; Ren, X.; Zhang, W. Electrospun scaffolds of silk fibroin and poly (lactide-co-glycolide) for endothelial cell growth. *J. Mater. Sci. Mater. Med.* **2015**, *26*, 1–14. [CrossRef]
95. Li, Y.-S.J.; Haga, J.H.; Chien, S. Molecular basis of the effects of shear stress on vascular endothelial cells. *J. Biomech.* **2005**, *38*, 1949–1971. [CrossRef]
96. Sessa, W.C. Enos at a glance. *J. Cell. Sci.* **2004**, *117*, 2427–2429. [CrossRef] [PubMed]
97. Kolluru, G.K.; Sinha, S.; Majumder, S.; Muley, A.; Siamwala, J.H.; Gupta, R.; Chatterjee, S. Shear stress promotes nitric oxide production in endothelial cells by sub-cellular delocalization of eNOS: A basis for shear stress mediated angiogenesis. *Nitric Oxide* **2010**, *22*, 304–315. [CrossRef] [PubMed]
98. Dimmeler, S.; Fleming, I.; Fisslthaler, B.; Hermann, C.; Busse, R.; Zeiher, A.M. Activation of nitric oxide synthase in endothelial cells by Akt-dependent phosphorylation. *Nature* **1999**, *399*, 601–605. [CrossRef] [PubMed]
99. Uematsu, M.; Ohara, Y.; Navas, J.P.; Nishida, K.; Murphy, T.; Alexander, R.W.; Nerem, R.M.; Harrison, D.G. Regulation of endothelial cell nitric oxide synthase mRNA expression by shear stress. *Am. J. Physiol.* **1995**, *269*, C1371–C1378. [CrossRef] [PubMed]
100. Buga, G.M.; Gold, M.E.; Fukuto, J.M.; Ignarro, L.J. Shear stress-induced release of nitric oxide from endothelial cells grown on beads. *Hypertension* **1991**, *17*, 187–193. [CrossRef] [PubMed]
101. Hsieh, H.-J.; Liu, C.-A.; Huang, B.; Tseng, A.H.; Wang, D.L. Shear-induced endothelial mechanotransduction: The interplay between reactive oxygen species (ROS) and nitric oxide (NO) and the pathophysiological implications. *J. Biomed. Sci.* **2014**, *21*, 1–15. [CrossRef] [PubMed]

Journal of
Clinical Medicine

MDPI

Article

The Future State of Newborn Stem Cell Banking

Katherine S. Brown [1,*], Mahendra S. Rao [2] and Heather L. Brown [1]

[1] Scientific and Medical Affairs, Cbr Systems Inc., 611 Gateway Blvd, Suite 820, South San Francisco, CA 94080, USA; hbrown@cordblood.com
[2] Collaborative Science Chair, InSTEM, Bangalore 560065, India; rao1789@gmail.com
* Correspondence: kbrown@cordblood.com; Tel.: +01-650-826-6179

Received: 21 December 2018; Accepted: 15 January 2019; Published: 18 January 2019

Abstract: Newborn stem cell banking began with the establishment of cord blood banks more than 25 years ago. Over the course of nearly three decades, there has been considerable evolution in the clinical application of stem cells isolated from newborn tissues. The industry now finds itself at an inflection point as personalized medicine and regenerative medicine continue to advance. In this review, we summarize our perspective on newborn stem cell banking in the context of the future potential that stem cells from perinatal tissues are likely to play in nascent applications. Specifically, we describe the relevance of newborn stem cell banking and how the cells stored can be utilized as starting material for the next generation of advanced cellular therapies and personalized medicine.

Keywords: stem cell banking; newborn stem cells; perinatal stem cells; umbilical cord tissue; umbilical cord blood; placenta

1. Current State of Newborn Stem Cell Banking

In September of 2018, the umbilical cord blood transplant and newborn stem cell banking communities celebrated the 30th anniversary of the first hematopoietic stem cell (HSC) transplant using cord blood as a graft for a patient with Fanconi's anemia. The successful demonstration that cord blood is capable of reconstituting a patient's blood and immune system, coupled with the confirmation that cord blood can be cryopreserved for later use, led to the establishment of cord blood banks, and thus the newborn stem cell banking industry, in the early 1990s [1]. Newborn stem cell banking encompasses public cord blood banks, which store cord blood units for use in an unrelated recipient; private banks, which store cord blood for future use by the donor or a first- or second-degree relative; and hybrid banks, which offer combined services [2]. It is estimated that more than more than 800,000 cord blood units are cryopreserved in public banks and over 5 million more are stored in private cord blood banks [3].

It is widely recognized that additional perinatal tissues routinely discarded as medical waste contain nonhematopoietic cells with potential therapeutic value. For example, mesenchymal stromal cells (MSCs) can be isolated from placental tissue, umbilical cord tissue, and amniotic fluid. With the exception of amniotic fluid, which is obtained during an elective amniocentesis, these tissues are collected in a noninvasive procedure following birth of the neonate and would otherwise be discarded. Concomitant cryopreservation of multiple newborn tissues from the same donor (see, for example, [4]) has been demonstrated. Based on the potential therapeutic value, enhanced proliferative capacity, lack of ethical controversies, and reduced risk of exposure to virus and environmental toxins of newborn stem cells compared to stem cells from adult tissues, numerous cord blood banks expanded their processes in order to cryopreserve additional tissues alongside umbilical cord blood, and in some instances, as a stand-alone product. These new cryopreservation products encompass umbilical cord tissue, placental tissue, amniotic fluid, and amniotic membrane. In addition to serving as a repository of MSCs, epithelial cells and progenitor cells, endothelial cells and progenitor cells, and subpopulations of cells that may have therapeutic value can be isolated from the aforementioned tissues. A summary of perinatal tissues that can be cryopreserved and representative cell populations obtained from each are provided in Figure 1. Those families storing cord blood at a private bank in the United States typically pay between $300 and $2300 for the collection, processing, and initial storage, with annual storage fees thereafter [1]. Storage of an additional newborn tissue, such as umbilical cord tissue or placental tissue, costs an additional $800–1300. There is no charge to families donating newborn tissues, as public banks cover costs associated with collection, processing, and storage. The Parent's Guide to Cord Blood Foundation [5] provides a global index of public and private banks and their respective services.

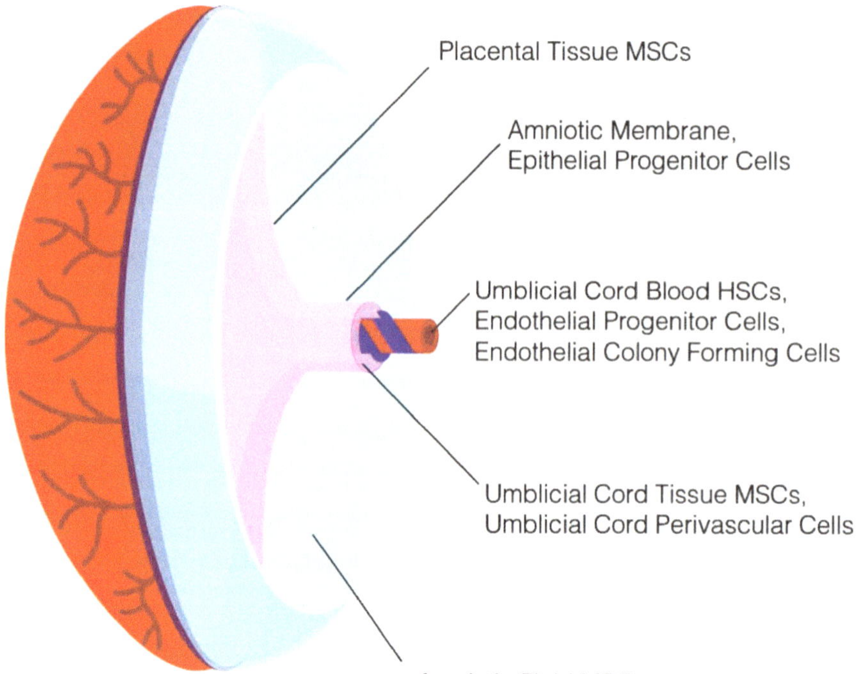

Figure 1. Schematic of newborn tissue that can be collected and stored for immediate or future use and the cell populations associated with each. Hematopoietic stem cells (HSCs) can be obtained from the umbilical cord blood, as can endothelial progenitor cells and endothelial colony-forming cells. Mesenchymal stem cells can be isolated from various locations within the placenta, umbilical cord tissue, amniotic membrane, and amniotic fluid. MSCs (mesenchymal stromal cells) can be obtained from umbilical cord blood, but successful isolation is time- and volume-dependent, rendering cord blood a less reliable source. The umbilical cord tissue is also a source of other stem or progenitor cells with potential applications.

Certain maternal and neonatal parameters associated with cord blood quality, such as gestational age and birth weight, can be used by public banks to optimize donor selection in an effort to increase likelihood of utilization and as part of managing costs associated with tissue procurement [6]. The cellular content of cord blood is also influenced by seasonal variation and circadian oscillations; consideration of time-related parameters of cord blood collection is proposed as a mechanism to target cord blood donations with greater hematopoietic potential [7]. While the hematopoietic potential of cord blood units can be estimated by determining the number of cells expressing the CD34 antigen (CD34+ cells), different strategies must be employed by newborn stem cell banks to determine the potency of MSCs from perinatal tissues. Expression of cell surface markers is routinely used to identify an MSC population, yet is a poor substitute for product characterization [8]. For example, the maternal metabolic environment has been reported to alter bioenergetic profile as well as expression of proteins involved in stress response, metabolic, and cytoskeletal pathways of MSCs, confirmed by immunophenotype, from perinatal tissues [9,10]. Donor-to-donor variability of perinatal MSCs in anti-inflammatory and immunomodulatory assays has also been reported [11,12]. These findings are consistent with observations of donor heterogeneity of bone marrow MSCs and underscore the importance of functional assessments for newborn stem cell banks. Public banks exploring storage of allogeneic MSCs from newborn tissue may incorporate screening of donor cell lines for desired characteristics, such as immunomodulatory or angiogenic properties, prior to cryopreservation. A practical approach for private banks is to evaluate the post-thaw MSC product utilizing an assay that provides an estimate of functionality within the context of the intended therapeutic application. Incorporation of functional assessments by private banks is a logical extension of comparability studies described in more detail below.

2. Newborn Stem Cells in Transplant and Regenerative Medicine Applications

More than 40,000 hematopoietic stem cell transplants using cord blood have been performed during the last three decades [13]. In this setting, the hematopoietic stem cells in the cord blood are utilized for homologous reconstitution of the blood and immune system in the same manner as a bone marrow transplant. Cord blood is recognized as an alternative graft source for hematopoietic stem cell transplant in pediatric and adult patients and has been used in the treatment of over 80 diseases, including hematologic malignancies and disorders, congenital immunodeficiency disorders, and certain metabolic disorders [14].

There is also considerable interest in exploring cord blood as a therapeutic intervention in nonhematopoietic indications. In the mid-2000s, researchers began investigating cord blood in acquired neurological indications. Pilot and clinical trials enrolling pediatric patients with conditions such as cerebral palsy, autism spectrum disorder, and acquired hearing loss have confirmed the safety, and in some patients evidence of efficacy, of administering minimally manipulated cord blood cryopreserved in an autologous setting [15–19]. A small phase I study also confirmed the safety and feasibility of administering allogeneic unrelated cord blood to adult ischemic stroke patients [20]. Based on the observed safety profile and preliminary evidence of efficacy, additional studies to determine efficacy and to evaluate the safety of the approach in human leukocyte antigen (HLA)-matched related and unrelated donor cord blood are either underway or planned [16,17,20].

A recent review of clinical trials employing perinatal tissue-derived products in advanced cell therapy identified 281 clinical studies registered between 2005 and 2015, and acquired neurological conditions or disorders was the second most common category of diagnosis behind trials in hematology or oncology with manipulated cell types [21]. Of more than 500 cord blood units released for clinical application from our institution, 80% have gone to clinical trials or experimental uses in regenerative medicine, with the vast majority of those indications being neurological injuries sustained at or around the time of birth or diagnosis associated with said injuries (Figure 2).

Cryopreserved perinatal tissues are also being explored for their capacity to augment established uses of cord blood in traditional transplant medicine. For example, monocytes isolated from cryopreserved cord blood are used to manufacture a cell therapy product aimed to augment cord blood transplantation in the setting of inherited demyelinating conditions of the central nervous system [22]. Furthermore, the potential for MSCs isolated from cryopreserved cord tissue or placental tissue to facilitate ex vivo expansion of cord blood hematopoietic stem cells has been reported and provides further rationale for storing multiple newborn tissues from the same donor.

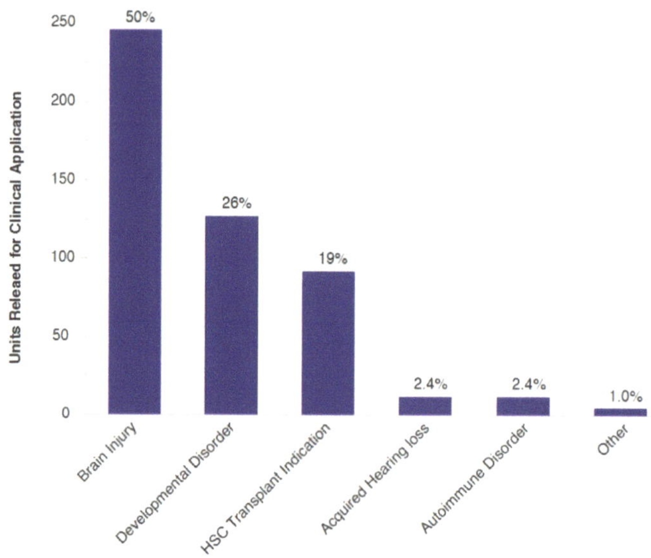

Figure 2. Categories of clinical applications for which cord blood units have been released from a private newborn stem cell bank. As of December 2018, the Cord Blood Registry has released over 500 cord blood units for use in clinical applications, slightly less than 20% of which were utilized in a hematopoietic stem cell transplant. Data are presented as the percentage of units released for each generalized category.

Interest in exploring cryopreserved newborn stem cells in regenerative applications has continued to increase over the past decade. The mechanism by which the therapeutic cells exert their effects in many of these exploratory studies is theorized to involve immunomodulation and paracrine-based signaling facilitating endogenous tissue repair rather than direct cell replacement or engraftment. The evolving landscape of clinical trials in regenerative medicine utilizing umbilical cord tissue, placental tissue, and minimally manipulated cord blood has been recently reviewed in depth by others, as have approaches to ex vivo cord blood expansion and efforts to improve outcomes following cord blood stem cell transplantation in adult recipients (see, for example, [2,13,23,24]). The remainder of this discussion will focus on the application of cryopreserved perinatal cells in more nascent technologies.

3. Emerging Advanced Cellular Therapies and Changes to the Business Model

Cord blood has established utility in the setting of hematopoietic stem cell transplant medicine, and cord blood stem cells are also being investigated for their ability to induce healing and repair tissue, which has the potential to greatly increase the utilization of cord blood in the clinical setting. Nevertheless, both private and public institutions within the newborn stem cell banking industry face mounting challenges.

Public cord blood banks have faced increased regulations over the past decade. In particular, in 2007, the FDA established that cord blood, unlike bone marrow, intended for use in an unrelated recipient is a processed, prescription product, or drug. Effective as of 2009, public banks are considered a "manufacturer" and are required to have approval from the FDA biologics license application (BLA) for cord blood. The public banking community has repeatedly pointed out that the process for achieving licensure is both onerous and costly, negatively impacting the costs of collecting, storing, and distributing cord blood units [25]. One factor influencing treating physicians is that many adult patients require two cord blood units to meet the cell dose thresholds for a hematopoietic stem cell transplant, adding complexity to the transplant itself and pushing cord blood towards being cost-prohibitive compared to other graft sources. To date, only seven public banks have successfully obtained licensure, while the remaining public banks are allowed to continue operations as they move towards compliance. Public banks recover costs when cord blood units are released from inventory, not at the time of cryopreservation. Additionally, public banks have a relatively low utilization rate, a significant contributor considering that nearly 90% of institutions have reported struggling financially [26]. Lastly, successful licensure is not retroactive for previously collected inventory; units collected pre-licensure, while theoretically of equivalent quality to licensed units, can only be used under an investigational new drug (IND) application, which are granted for specific uses.

With a positive safety profile to date, the use of cord blood cells in regenerative medicine applications appears poised to increase the number of clinical settings in which the cells can be considered as part of a therapeutic intervention, as discussed above. This rapid evolution for potential indications outside of hematopoietic reconstitution has the potential to greatly influence the utilization rate of cord blood units from the public inventory. Each bank, though, would need to be approved for releasing cord blood units for new indications by performing the requisite clinical studies, which public banks have neither the resources nor the commercial initiative to do. Private cord blood banks are indirectly affected by this issue due to a lack of comparability studies. To our knowledge, only one controlled study in a regenerative medicine application has directly compared infusion of autologous cord blood to allogeneic unrelated cord blood from a public cord blood bank [27]. Should allogeneic unrelated cord blood prove to be therapeutically relevant in a regenerative medicine setting, for example, patients with autism spectrum disorder (ASD), a public bank could consider applying for a BLA for use in ASD, which would be separate from the BLA for uses in transplant medicine. This process would establish for the public banks another avenue for cord blood unit utilization and revenue upon unit release. For private cord blood banks, where revenue is recognized at the time of storage rather than release, there is less motivation to invest in clinical trials as cord blood units could be released under an IND held by the treating facility. The dilemma then is that private institutions, which are best suited to commit financially to exploring new indications, have the least motivation to do so from the perspective of short-term revenue generation.

Commercial institutions in adjacent business areas are likely to influence the near-term financial sustainability of public cord blood banks. Companies such as Gamida Cell Ltd, Fate Therapeutics, and others are focused on expansion technologies and approaches for improving the efficiency of homing and engraftment of cord blood stem cells. While applicable to both private and publicly banked units, public banks are likely to benefit most from successful clinical translation if these technologies increase utilization of units that would otherwise fail to meet cell dose thresholds. Nohla Therapeutics is taking a different approach by developing off-the-shelf, ex vivo expanded products from cord blood units to provide a short-term hematopoietic bridge following transplant of an unmanipulated cord blood graft. If successful, Nohla's product could help improve outcomes in patients undergoing a cord blood transplant. These technologies address current challenges in cord blood hematopoietic stem cell transplantation and their success could increase utilization rate from the public inventory, which in turn would help bolster the financial health of public banks.

Public cord blood banks also have the opportunity to diversify revenue streams through creative outsourcing of inventory that is unlikely to be used as a graft for hematopoietic reconstitution. Blood banks collecting and distributing peripheral blood for use in transfusion medicine have already begun a similar shift as they look to capitalize on ancillary material or products that have expired. Likewise, the increasing demand for human platelet lysate as a supplement for use in routine cell culture has led to a niche market for blood banks. Public cord blood banks are exploring the applicability of donated cord blood in nascent technologies, such as chimeric antigen receptor T-cell (CAR-T) products. Importantly, a single cord blood unit is often sufficient for CAR-T, and for these and other similar immunotherapies, the cells are only required for a short time. Because units are HLA-typed at the time of processing, there is also a great deal of interest in utilizing inventory from public cord blood banks to create a haplobank of HLA-homozygous induced pluripotent stem cells (iPSCs); proof-of-concept studies have demonstrated the feasibility of this approach for populations in various geographical locations [28].

Private newborn stem cell banks are also interested in exploring the application of cord blood in the evolving field of immunotherapies and more than minimally manipulated products. Clinical studies initiated in the mid-2000s sought to determine the potential of nonmanipulated cryopreserved cord blood in Type 1 diabetes. Consistent with the results of clinical trials in acquired neurological injuries, autologous infusion of cord blood in patients with Type 1 diabetes was confirmed to be safe and feasible both alone and when administered alongside daily vitamin D and docosahexaenoic acid [29,30]. Investigators postulated that the limited number of regulatory T cells in cord blood limited the potential for sustained preservation of C-peptide, raising the possibility that infusion of regulatory T cells isolated and expanded from cord blood may be more efficacious than the heterogeneous cell populations in nonmanipulated cord blood [30]. The feasibility of expanding regulatory T cells from privately banked cord blood in a current good manufacturing practice (cGMP) setting was recently confirmed, opening the door for the approach to be explored in a clinical trial setting for Type 1 diabetes and other autoimmune conditions [31]. It is also worth noting that with the refinement in the safety and efficiency of gene editing capabilities and early successes in gene therapy clinical trials, one can foresee cord blood collection and banking in a private setting with the future intent of personalized, autologous gene therapy for individuals with a known genetic disorder in the future. Private banks have also shown interest in leveraging the technical advancements in iPSC reprogramming on behalf of their clients. In a private bank setting, starting material can be used to generate personalized, donor-specific iPSCs for autologous use. In fact, cells from cord blood and cord tissue from the same donor stored at a private bank can be utilized to generate iPSCs; confirmation that lines generated from either source material are of equivalent quality provided rationale for utilizing the cord tissue cells as starting material, preserving the cord blood unit in its entirety for future clinical utility [32]. Additionally, MSCs isolated from previously cryopreserved cord tissue at a private bank are amenable to reprogramming with multiple integration free methods on semi- or fully automated technology platforms for enhanced standardization and scalability [33,34]. One potential business model would be to utilize a portion of collected newborn material, either cord blood or cord tissue, to generate a biologically potent, individualized iPSC line which is then stored as a companion product for future potential uses. Extracellular vesicles, including exosomes, represent another intriguing potential companion storage offering. A number of commercial institutions have established off-the-shelf stem cell-derived extracellular vesicle products and are moving them into clinical trials. It will be interesting to see what influence, if any, the burgeoning field of cell-free therapeutics has on the newborn stem cell banking industry.

There is increased clarity on potential application of privately banked stem cells outside of the established uses in HSC transplants. Early recognition of the interest and anticipated utilization of MSCs across a variety of clinical settings garnered the interest of the private banks as an opportunity to leverage infrastructure and technology platforms to provide storage of umbilical cord tissue as a service under the collection and manufacturing model already established as part of private cord blood banking. Similarly, private institutions led the industry in the banking of placental tissue. Although efforts are underway (see, for example, [35]), public banks have been slower to establish the same programs, as the path for and extent of reimbursement will remain undefined until there are established and more widely practiced clinical applications for cells derived from these alternative perinatal tissues.

For cord blood, clinical outcomes of transplants are influenced by the graft characteristics, including nucleated cell dose, stem cell dose, and HLA match, while the impact of the volume reduction processing technology used in preparing cord blood for cryopreservation is less evident, assuming that units selected meet appropriate criteria for the individual recipient and appropriate methods for thawing are followed [36,37]. While cryopreservation of umbilical cord tissue and placenta as source material have been an integral part of stimulating new avenues of clinical research, their rapid adoption led to a variety of different approaches for preparation and cryopreservation. Thus, private newborn stem cell banks and some public banks find themselves in a position where similar material is stored, but comparability of the final product has yet to be determined. The industry clearly finds itself at a point where comparability studies, to confirm that products processed and stored via any number of processes, are increasingly warranted. Cord tissue and placental tissue can be prepared for cryopreservation as either a cell suspension or composite material thawed at a later date for isolation of cells. Thus, assays for functional attributes of the final product after thawing of a cell suspension, or in the case of whole tissue cryopreservation, recovery of cells from thawed composite material, are a logical point for implementing a standardized approach to determine comparability. Importantly, assessments should ideally encompass functional attributes associated with mechanisms of action, rather than simply identity and purity of the cell population. These studies require investment from private institutions, but are well justified given the benefit to the client as well as scientific and medical communities.

4. Changing Landscape of the Newborn Stem Cell Banking Market

The proliferation of private cord blood banks in the early 2000s led to market saturation in many geographical regions. Mergers and acquisitions subsequently led to consolidation in the private banking industry. It is estimated that over the last decade, the number of European banks decreased by one-third through consolidation activity [38]. The merger of Cell Care Australia, the largest private newborn bank in Australia, with Insception Lifebank, the largest private bank in Canada, exemplify industry willingness to explore major intercontinental activity. Although a relatively late arrival to the newborn stem cell banking landscape, India is poised to overtake the market for cord blood banking and families have the opportunity to choose from any number of different banks. One of those options is Cryo-Save, which leverages processing capabilities and economies of scale by operating only several centralized processing facilities in Europe, India, and South Africa, with numerous regional facilities operating under the Cryo-Save trademark through licensing agreements. In stark contrast, the government of China allows only a single cord blood bank to operate in each province; although consumer choice is limited, each licensed bank must function as a hybrid bank, providing donation-based and private storage services.

Hybrid banks were originally met with resistance from the banking community based on perceived conflicts of interest for the donor. More recently, there has been greater acceptance of the hybrid model, with several institutions demonstrating that private banking can be used to offset costs for altruistic public donations without deterring from the donor pool. Companies such as StemCyte International have found stability in the consumer market while also providing donated cord blood units for use in unrelated recipients. As evidenced by the acquisition of the hybrid bank CORD:USE by the private bank Cryo-Cell International, mergers and acquisitions also present an opportunity to enter into new market sectors, such as public banking, while limiting the potential financial risks of de novo development. Some industry activities meanwhile capitalize on infrastructure expertise and capacity for biobanking as a logical extension of services. For example, Cord Blood Registry was acquired and merged with California Cryobank. The newly established California Cryobank Life Sciences Platform provides newborn stem cell banking alongside reproductive tissue services. Celularity Inc., a spinout of Celgene, is developing placenta-derived allogeneic immuno-oncology and regenerative products while offering cryopreservation of umbilical cord and placental tissues through the private arm LifeBank USA. Celularity also acquired CariCord, a private cord blood bank affiliated with ClinImmune laboratories, presumably in part to augment biosourcing capabilities for future placenta-derived products.

5. Conclusions

Within the United States, the health care system is critical to long-term economic stability for newborn stem cell banking. Health care coverage of the costs associated with collection and cryopreservation of newborn tissues for future use and establishing insurance reimbursement once clinical efficacy and cost calculations are established in regenerative medicine applications would provide for economic incentives for all invested parties and help the industry meet the increasing demand for precision health care. Outside of establishing legislation mandating education by health care providers for expecting families, efforts to lobby Congress have thus far been unsuccessful. The Cord Blood Association, a nonprofit organization comprised of stakeholders across the banking industry, advocates on behalf of the community to advance relevant legislation and modifications to the regulatory framework and may be a more effective approach than previous efforts.

Emerging technologies have the potential to influence the direction of newborn stem cell banking; both public and private banking institutions will need to identify a strategic path in order to position themselves favorably for the long term. Public and private banks are actively exploring ways to augment their model from providing storage of a minimally manipulated cellular product to one that recognizes the promise in companion products and provides starting material for downstream applications (Figure 3). It is also clear that private and public cord blood banks will need to react differently to the financial and industry challenges, based on the divergence in their models and how regulations are applied to each.

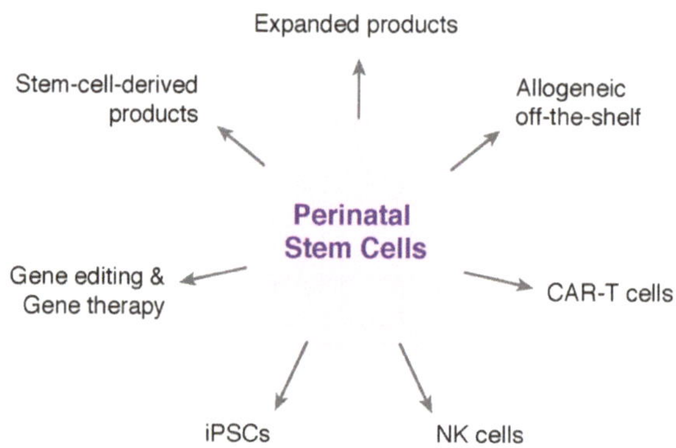

Figure 3. Newborn stem cells in emerging and advanced cellular therapies. Stem or progenitor cells obtained from various newborn tissues are depicted in the center, while potential downstream products are represented on the periphery. iPSCs, induced pluripotent stem cells; NK cells, natural killer cells; CAR-T cells, chimeric antigen receptor T cells.

Author Contributions: Conceptualization, K.S.B. and M.S.R. Writing-original draft preparation, K.S.B. Writing-review and editing, K.S.B., M.S.R., and H.L.B.

Acknowledgments: The authors would like to thank Peter Lundblad for assistance with preparation of the figures.

Conflicts of Interest: Cbr Systems, Inc. (Cord Blood Registry) is part of the California Cryobank Lifescience Platform K.B. and H.B. receive payment as employees of the Cord Blood Registry. M.R. receives compensation as a member of the Cord Blood Registry's scientific advisory board.

References

1. Ballen, K.K.; Verter, F.; Kurtzberg, J. Umbilical cord blood donation: Public or private? *Bone Marrow Transpl.* **2015**, *50*, 1271–1278. [CrossRef]
2. Dessels, C.; Alessandrini, M.; Pepper, M.S. Factors Influencing the Umbilical Cord Blood Stem Cell Industry: An Evolving Treatment Landscape. *Stem Cells Transl. Med.* **2018**, *7*, 643–650. [CrossRef]
3. Kurtzberg, J. A History of Cord Blood Banking and Transplantation. *Stem Cells Transl. Med.* **2017**, *6*, 1309–1311. [CrossRef] [PubMed]

4. Skiles, M.L.; Brown, K.S.; Tatz, W.; Swingle, K.; Brown, H.L. Quantitative analysis of composite umbilical cord tissue health using a standardized explant approach and an assay of metabolic activity. *Cytotherapy* **2018**, *20*, 564–575. [CrossRef] [PubMed]
5. The Parent's Guide to Cord Blood Foundation. Available online: http://www.parentsguidecordblood.org (accessed on 15 December 2018).
6. Page, K.M.; Mendizabal, A.; Betz-Stablein, B.; Wease, S.; Shoulars, K.; Gentry, T.; Prasad, V.K.; Sun, J.; Carter, S.; Balber, A.E.; et al. Optimizing donor selection for public cord blood banking: Influence of maternal, infant, and collection characteristics on cord blood unit quality. *Transfusion* **2013**, *54*, 340–352. [CrossRef] [PubMed]
7. Mazzoccoli, G.; Miscio, G.; Fontana, A.; Copetti, M.; Francavilla, M.; Bosi, A.; Perfetto, F.; Valoriani, A.; De Cata, A.; Santodirocco, M.; et al. Time related variations in stem cell harvesting of umbilical cord blood. *Sci. Rep.* **2016**, *6*, 21404. [CrossRef] [PubMed]
8. Le Blanc, K.; Davies, L.C. MSCs-cells with many sides. *Cytotherapy* **2018**, *20*, 273–278. [CrossRef]
9. Capobianco, V.; Caterino, M.; Iaffaldano, L.; Nardelli, C.; Sirico, A.; Del Vecchio, L.; Martinelli, P.; Pastore, L.; Sacchetti, L. Proteome analysis of human amniotic mesenchymal stem cells (hA-MSCs) reveals impaired antioxidant ability, cytoskeleton and metabolic functionality in maternal obesity. *Sci. Rep.* **2016**, *6*, 25270. [CrossRef]
10. Iaffaldano, L.; Nardelli, C.; D'Alessio, F.; D'Argenio, V.; Nunziato, M.; Mauriello, L.; Procaccini, C.; Maruotti, G.M.; Martinelli, P.; Matarese, G.; et al. Altered Bioenergetic Profile in Umbilical Cord and Amniotic Mesenchymal Stem Cells from Newborns of Obese Women. *Stem Cells Dev.* **2018**, *27*, 199–206. [CrossRef]
11. Sutton, M.T.; Kaur, S.; Brown, K.S.; Skiles, M.L.; Folz, M.A.; Caplan, A.I.; Bonfield, T.L. Anti-inflammatory therapeutic development and optimization of umbilical cord tissue derived mesenchymal stem cells. *J. Stem Cell Res. Ther.* **2018**, *8*, 435. [CrossRef]
12. Peltzer, J.; Montespan, F.; Thepenier, C.; Boutin, L.; Uzan, G.; Rouas-Freiss, N.; Lataillade, J.-J. Heterogeneous functions of perinatal mesenchymal stromal cells require a preselection before their banking for clinical use. *Stem Cells Dev.* **2014**, *24*, 329–344. [CrossRef] [PubMed]
13. Ballen, K. Update on umbilical cord blood transplantation. *F1000Research* **2017**, *6*, 1556. [CrossRef] [PubMed]
14. Cairo, M.S.; Kindwall-Keller, T.L. Indications for Utilization of Stored Umbilical Cord Blood. Available online: https://www.cb-association.org/indications-for-utilization-of-stored-umbilical-cord-blood (accessed on 15 December 2018).
15. Baumgartner, L.S.; Moore, E.; Shook, D.; Messina, S.; Day, M.C.; Green, J.; Nandy, R.; Seidman, M.; Baumgartner, J.E. Safety of Autologous Umbilical Cord Blood Therapy for Acquired Sensorineural Hearing Loss in Children. *J. Audiol. Otol.* **2018**, *22*, 209–222. [CrossRef]
16. Dawson, G.; Sun, J.M.; Davlantis, K.S.; Murias, M.; Franz, L.; Troy, J.; Simmons, R.; Sabatos-DeVito, M.; Durham, R.; Kurtzberg, J. Autologous Cord Blood Infusions Are Safe and Feasible in Young Children with Autism Spectrum Disorder: Results of a Single-Center Phase I Open-Label Trial. *Stem Cells Transl. Med.* **2017**, *6*, 1332–1339. [CrossRef]
17. Sun, J.M.; Song, A.W.; Case, L.E.; Mikati, M.A.; Gustafson, K.E.; Simmons, R.; Goldstein, R.; Petry, J.; McLaughlin, C.; Waters-Pick, B.; et al. Effect of Autologous Cord Blood Infusion on Motor Function and Brain Connectivity in Young Children with Cerebral Palsy: A Randomized, Placebo-Controlled Trial. *Stem Cells Transl. Med.* **2017**, *6*, 2071–2078. [CrossRef] [PubMed]
18. Chez, M.; Lepage, C.; Parise, C.; Dang-Chu, A.; Hankins, A.; Carroll, M. Safety and Observations from a Placebo-Controlled, Crossover Study to Assess Use of Autologous Umbilical Cord Blood Stem Cells to Improve Symptoms in Children with Autism. *Stem Cells Transl. Med.* **2018**, *7*, 333–341. [CrossRef] [PubMed]
19. Sun, J.; Allison, J.; McLaughlin, C.; Sledge, L.; Waters-Pick, B.; Wease, S.; Kurtzberg, J. Differences in quality between privately and publicly banked umbilical cord blood units: A pilot study of autologous cord blood infusion in children with acquired neurologic disorders. *Transfusion* **2010**, *50*, 1980–1987. [CrossRef]
20. Laskowitz, D.T.; Bennett, E.R.; Durham, R.J.; Volpi, J.J.; Wiese, J.R.; Frankel, M.; Shpall, E.; Wilson, J.M.; Troy, J.; Kurtzberg, J. Allogeneic Umbilical Cord Blood Infusion for Adults with Ischemic Stroke: Clinical Outcomes from a Phase 1 Safety Study. *Stem Cells Transl. Med.* **2018**, *7*, 521–529. [CrossRef]
21. Couto, P.S.; Bersenev, A.; Verter, F. The first decade of advanced cell therapy clinical trials using perinatal cells (2005–2015). *Regen. Med.* **2017**, *12*, 953–968. [CrossRef]

22. Saha, A.; Buntz, S.; Scotland, P.; Xu, L.; Noeldner, P.; Patel, S.; Wollish, A.; Gunaratne, A.; Gentry, T.; Troy, J.; et al. A cord blood monocyte-derived cell therapy product accelerates brain remyelination. *JCI Insight* **2016**, *1*, e86667. [CrossRef]
23. Becnel, M.; Shpall, E.J. Current and future status of stem cell expansion. *Curr. Opin. Hematol.* **2018**, *25*, 446–451. [CrossRef]
24. Abbaspanah, B.; Momeni, M.; Ebrahimi, M.; Mousavi, S.H. Advances in perinatal stem cells research: A precious cell source for clinical applications. *Regen. Med.* **2018**, *13*, 595–610. [CrossRef] [PubMed]
25. Kapinos, K.A.; Whaley, C. *Challenges to the Sustainability of the U.S. Public Cord Blood System*; RAND Corporation: Santa monica, CA, USA, 2017.
26. Magalon, J.; Maiers, M.; Kurtzberg, J.; Navarrete, C.; Rubinstein, P.; Brown, C.; Schramm, C.; Larghero, J.; Katsahian, S.; Chabannon, C.; et al. Banking or Bankrupting: Strategies for Sustaining the Economic Future of Public Cord Blood Banks. *PLoS ONE* **2015**, *10*, e0143440. [CrossRef] [PubMed]
27. Kurtzberg, J. Cord Blood Infusion for Children with Autism Spectrum Disorder (Duke ACT). Available online: https://clinicaltrialsgov/ct2/show/NCT02847182 (accessed on 15 December 2018).
28. Lee, S.; Huh, J.Y.; Turner, D.M.; Lee, S.; Robinson, J.; Stein, J.E.; Shim, S.H.; Hong, C.P.; Kang, M.S.; Nakagawa, M.; et al. Repurposing the Cord Blood Bank for Haplobanking of HLA-Homozygous iPSCs and Their Usefulness to Multiple Populations. *Stem Cells* **2018**, *36*, 1552–1566. [CrossRef]
29. Haller, M.J.; Wasserfall, C.H.; Hulme, M.A.; Cintron, M.; Brusko, T.M.; McGrail, K.M.; Sumrall, T.M.; Wingard, J.R.; Theriaque, D.W.; Shuster, J.J.; et al. Autologous umbilical cord blood transfusion in young children with type 1 diabetes fails to preserve C-peptide. *Diabetes Care* **2011**, *34*, 2567–2569. [CrossRef] [PubMed]
30. Haller, M.J.; Wasserfall, C.H.; Hulme, M.A.; Cintron, M.; Brusko, T.M.; McGrail, K.M.; Wingard, J.R.; Theriaque, D.W.; Shuster, J.J.; Ferguson, R.J.; et al. Autologous umbilical cord blood infusion followed by oral docosahexaenoic acid and vitamin D supplementation for C-peptide preservation in children with Type 1 diabetes. *Biol. Blood Marrow Transpl.* **2013**, *19*, 1126–1129. [CrossRef]
31. Seay, H.R.; Putnam, A.L.; Cserny, J.; Posgai, A.L.; Rosenau, E.H.; Wingard, J.R.; Girard, K.F.; Kraus, M.; Lares, A.P.; Brown, H.L.; et al. Expansion of Human Tregs from Cryopreserved Umbilical Cord Blood for GMP-Compliant Autologous Adoptive Cell Transfer Therapy. *Mol. Ther. Methods Clin. Dev.* **2016**, *4*, 178–191. [CrossRef]
32. Mack, A.A.; Brown, K.S.; Faust, E.; Monroe, K.; Wachowiak, R.; Wang, W.B.; Skiles, M.L. GMP-compatible iPSC derivation from multiple perinatal tissue sources from the same donor. In Proceedings of the Annual Meeting of the International Society for Stem Cell Research, Boston, MA, USA, 14–17 June 2017.
33. Paull, D.; Sevilla, A.; Zhou, H.; Hahn, A.K.; Kim, H.; Napolitano, C.; Tsankov, A.; Shang, L.; Krumholz, K.; Jagadeesan, P.; et al. Automated, high-throughput derivation, characterization and differentiation of induced pluripotent stem cells. *Nat. Methods* **2015**, *12*, 885–892. [CrossRef]
34. Hunter, C.; McCarthy, B.; Kroeker, T.; Reggio, K.; Zhou, M.; Jagadeesan, P.; Brown, K.S.; Sun, B.; Moroziewicz, D.; Nadar, V.; et al. Generation of iPSCs from umbilical cord tissue using multiple integration free reprograming methods. In Proceedings of the AABB Annual Meeting, Boston, MA, USA, 13–16 October 2018.
35. Emnett, R.J.; Kaul, A.; Babic, A.; Geiler, V.; Regan, D. Evaluation of tissue homogenization to support the generation of GMP-compliant mesenchymal stromal cells from the umbilical cord. *Stem Cells Int.* **2016**, *2016*, 3274054. [CrossRef]
36. Nikiforow, S.; Li, S.; Snow, K.; Liney, D.; Kao, G.S.-H.; Haspel, R.; Shpall, E.J.; Glotzbecker, B.; Sica, R.A.; Armand, P.; et al. Lack of impact of umbilical cord blood unit processing techniques on clinical outcomes in adult double cord blood transplant recipients. *Cytotherapy* **2016**, *19*, 272–284. [CrossRef]
37. Ballen, K.K.; Logan, B.R.; Laughlin, M.J.; He, W.; Ambruso, D.R.; Armitage, S.E.; Beddard, R.L.; Bhatla, D.; Hwang, W.Y.K.; Kiss, J.E.; et al. Effect of cord blood processing on transplantation outcomes after single myeloablative umbilical cord blood transplantation. *Biol. Blood Marrow Transpl.* **2014**, *21*, 688–695. [CrossRef] [PubMed]
38. Hildreth, C. Cord Blood Banking Market Consolidation. Available online: https://bioinformant.com/cord-blood-banking-market-consolidation (accessed on 11 December 2018).

Journal of
Clinical Medicine

MDPI

Review

Biobanking Organoids or Ground-State Stem Cells?

Wa Xian [1,2,*], Marcin Duleba [3], Yusuke Yamamoto [4], Matthew Vincent [5] and Frank McKeon [3]

1 Institute of Molecular Medicine, McGovern Medical School of University of Texas Health Science Center, Houston, TX 77030, USA
2 Department of Biochemistry and Molecular Biology, University of Texas McGovern Medical School, Houston, TX 77030, USA
3 Department of Biology and Biochemistry, University of Houston, Houston, TX 77204, USA; marcin.duleba@gmail.com (M.D.); fdmckeon@uh.edu (F.M.)
4 Division of Molecular and Cellular Medicine, National Cancer Center Research Institute, Tokyo 1040045, Japan; yusukeyamamoto198176@gmail.com
5 Tract Pharmaceuticals, Inc., Marlborough, MA 01752, USA; mvincent@tractpharma.com
* Correspondence: wa.xian@uth.tmc.edu; Tel.: +1-860-480-1188

Received: 12 November 2018; Accepted: 13 December 2018; Published: 16 December 2018

Abstract: Autologous transplantation of human epidermal stem cells cultured in Green's method is one of the first examples of utilizing adult stem cells in regenerative medicine. Using the same method, we cloned p63-expressing distal airway stem cells and showed their essential role in lung regeneration in a mouse model of acute respiratory distress syndrome. However, adult stem cells of columnar epithelial tissues had until recently evaded all attempts at cloning. To address this issue, we developed a novel technology that enabled cloning ground-state stem cells of the columnar epithelium. The adaption of this technology to clone stem cells of cancer precursors furthered our understanding of the dynamics of processes such as clonal evolution and dominance in Barrett's esophagus, as well as for testing platforms for chemical screening. Taken together, the properties of these ground-state stem cells, including unlimited propagation, genomic stability, and regio-specificity, make them ideal for regenerative medicine, disease modeling and drug discovery.

Keywords: adult stem cells; ground-state; autologous transplantation; regenerative medicine; biobanking

---

1. Introduction

One of the most exciting directions in medicine is based on stem cell research, which holds the promise of treatments and cures for various diseases and conditions that have so far frustrated traditional pharmaceutical approaches. Both stem cell research and clinical trials in regenerative medicine are presently dominated by pluripotent stem cells: embryonic stem cells (ESC) and induced pluripotent stem cells (iPSCs). However, these remarkable cells face important and unresolved hurdles, including the risk of generating teratomas, the truly arduous and inefficient processes of directed commitment to desired lineages and the limited regenerative capacity of derived lineages [1–6]. The promise of pluripotent stem cells has largely overshadowed efforts to harness so-called "adult" or "somatic" stem cells intrinsic to regenerative tissues. Green and colleagues developed methods for cloning epidermal stem cells [7] in their most immature and clonogenic form, which could be differentiated at will to yield a stratified squamous epithelium. The Green strategy has also been applied to capture stem cells from other stratified epithelia, including those of the corneal, thymic and airway epithelia [8–10]. However, stem cells of columnar epithelial tissues, such as those of the gastrointestinal tract, liver and kidney, have resisted cloning in a manner that maintains their immaturity during proliferative expansion. Instead, many researchers maintain these columnar epithelia as regenerative "organoids" that contain a minor fraction of stem cells that drive their growth [11–13]. While their potential for regenerative medicine is obvious [14], organoids are labor-intensive, very slow to expand and mostly comprised of differentiated cells rather than regenerative stem cells. To overcome this significant barrier in adult stem cell biology, highly robust technologies have been developed to clone and propagate clonogenic, "ground-state" stem cells of the human intestine and colon [15]. These cultured stem cells show remarkable stability in their genomic integrity and epigenetic commitment programs, show unlimited replicative expansion and maintain high clonogenicity, suggesting tremendous potential in disease modeling and regenerative medicine.

2. Technologies for Adult Epithelial Stem Cell Culturing

Since 1975, the 3T3J2 feeder system generated in Green's lab has been used widely to culture adult stem cells derived from p63-expressing stratified epithelium, including skin, thymus and lung [7,9,10]. Recently, a new technology developed in our lab, using a combination of 3T3J2 feeder cells and a specialized medium, was employed to clone ground-state stem cells from the columnar epithelium [15–17]. The common feature of these two feeder-dependent systems is their ability to keep highly clonogenic adult stem cells in culture so that stem cells can rapidly proliferate and reach sufficient numbers required for downstream applications, such as autologous transplantation, in a short time. In addition, the adult stem cells cultured in these feeder systems maintain the unipotent (e.g., skin) or multipotent differentiation ability (e.g., lung and intestine) and region-specificity despite long-term culturing [7,10,15,18]. The J2 strain of 3T3 cells, originally developed by Green, was used in both of these systems and cleared regulatory hurdles for clinical use thirty years ago, without reports of adverse effects [19–24]. Several clinical-grade cell culture methods have been developed under feeder-free conditions [20–22]. However, methods based on the Rheinwald and Green culture protocol remain the gold standard for clinical applications [23] due to the high clonogenicity and regenerative capacity.

Due to the different signaling requirements of the maintenance of stemness in the stratified epithelium and columnar epithelium, the growth medium of Green's method and the method that we developed (herein Xian–McKeon method) are significantly distinct. In addition to the growth factors such as IGF and EGF that were included in the medium in Green's method, the Xian–McKeon lab developed a media containing novel combinations of growth factors and regulators of TGF-β, Wnt/β-catenin, EGF, IGF, and Notch pathways [25,26] that supports the maintenance of human columnar epithelial stem cells, including intestinal stem cells, in a highly clonogenic, ground-state form. Importantly, single-cell derived, pedigree lines of human intestinal stem cells can be induced to differentiate into all cell types of the intestinal epithelium in air–liquid interface culture systems (Figure 1).

Regio-specificity of adult stem cells has been demonstrated in both of these systems [10,15]. In the Xian–McKeon method, the ground-state stem cells possessed robust epigenetic programs of commitment to regio-specific intestinal differentiation that are stable, despite more than a year of continuous propagation. This cell-autonomous regio-specificity of stem cells along the intestinal tract argues against a unitary "intestinal stem cell" or even one for each of the histologically recognized segments, and rather demonstrates a developmentally established spectrum of stem cells that ultimately maintain the histological and functional properties that define these segments. A heuristic deciphering of the commitment code from the regio-specific expression patterns described in Wang et al. [15] will guide parallel efforts with iPSCs to achieve appropriate lineage fates.

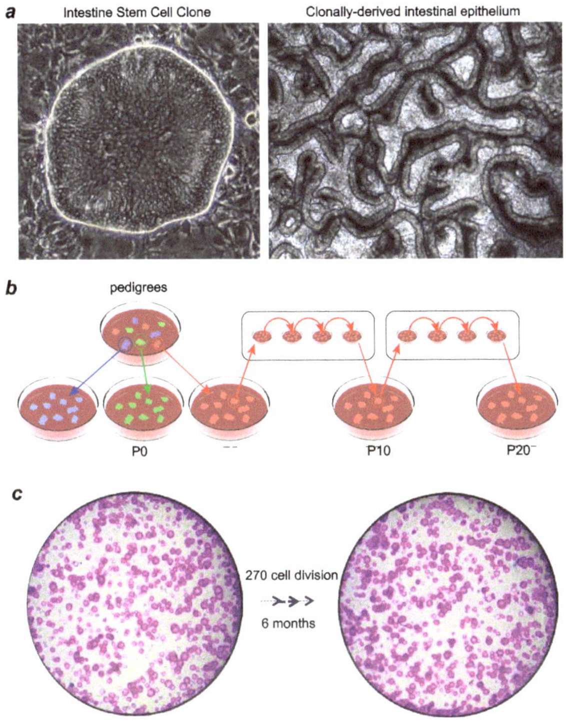

Figure 1. Xian–McKeon method of cloning ground-state intestinal stem cells. (**a**) Right, Representative image of a single-cell derived ISC^{GS} colony. Left, Top view of in vitro intestinal epithelium generated from stem cells of a pedigree of ISC^{GS} differentiated in an air–liquid interface setting. (**b**) Schematic diagram of establishing pedigrees from ISC^{GS}. Long-term culturing does not affect the capacity of ISC^{GS} to form single-cell derived pedigrees. (**c**) Clonogenicity assay revealing nearly unchanged number of Rhodamine red-stained colonies despite long-term culturing.

Despite the extensive studies on p63-expressing stem cells in the upper airways [27] and our own findings of p63-expressing cells contributing to lung regeneration following acute respiratory distress syndrome in mice [10,18], it was completely unclear whether we were dealing with one p63+ airway stem cell occupying different niches or many that were committed to regio-specific differentiation. Green's method was used to clone tracheal airway stem cells (TASCs) and distal airway stem cells (DASCs). These regio-specific airway stem cells demonstrated a clear distinction in their differentiation ability in various in vitro assays and in vivo transplantation experiments. Thus, despite the morphological and gene expression similarities between the upper airway stem cells and the distal airway stem cells, they clearly possess distinct gene expression profiles and cell fate programs. However, in the final analysis, the identification of distinct stem cells in the airways is of critical importance for any future scheme in regenerative medicine or the testing of drugs or biologics for their ability to rally such cells for regenerative endpoints.

Human ESC and iPSC lines acquire, with successive passages, genomic structural variations, including some that confer a selective advantage [28,29]. To assess the genomic stability of ground-state stem cells, we examined the copy number (CNV) and single nucleotide variation (SNV) in our cloned ground-state intestinal stem cell pedigrees and showed that ground-state stem cells sustain few genomic changes within the first 100 days of proliferative expansion [15]. Due to their high clonogenicity, we estimated that more than one billion stem cells would be generated from a single stem cell following 60 days culturing in this system [17]. Previous research suggested that epithelial cells require the loss of the pRB/p16^{INK4a} cell cycle control mechanisms, in addition to hTERT overexpression, to achieve immortality [30]. However, the remarkable proliferative potential of these ground-state stem cells, despite the absence of RB or p16 mutations suggests that their apparent immortality is a normal and intrinsic property of adult stem cells, rather than an indication of pathology.

3. Ground-State Stem Cells versus Organoids

Mammalian cells have been cultured in collagen and laminin-rich matrices as organoids in 3D culture since at least 1980 [12,31–35]. An organoid is defined as a unit of function of a given organ that is able to reproduce, in vitro, a biological structure similar in architecture and function to its counterpart in vivo. There are multiple origins of organoids that include a fragment of tissue, a stem cell isolated in an adult organ, an embryonic stem cell, or an induced pluripotent stem cell. Importantly, organoids can also be generated from cells that have lost stem cell capacity—so-called "transit amplifying cells" with limited proliferative potential—indicating that organoids per se are not strictly assays for stem cells [36].

Despite fundamental advances in stem cells afforded by organoid technology [37], the organoid method does not support cloning the intestinal stem cells or any other adult stem cells in their highly immature, clonogenic state. We solved the problem of culturing populations of pure stem cells of the human gastrointestinal tract using a cocktail of factors impacting Wnt, Notch, and TGF-β signaling pathways among others and were able to clone ground-state stem cells from all regions of the intestine and colon from endoscopic biopsies [15,17]. In other words, the stem cells of columnar epithelia can now be captured in the same ground-state that Green showed for the stem cells of stratified epithelia such as the epidermis. We found the following properties of these ground-state stem cells that make this technology particularly advantageous compared with the organoid method: (1) a typical 1 mm endoscopic biopsy yields 100–300 independent stem cell clones; (2) each of these clones can be independently propagated while maintaining genomic integrity; (3) these clones uniformly express stem cell markers and have a clonogenicity rate of 60–90%; (4) each of these clones retain epigenetic memory of the region from which they were derived; and (5) single clones can be differentiated to yield *3-D* intestinal epithelia with regionally-appropriate cell types.

It should be acknowledged that many of the features of these gastrointestinal stem cells were unknown prior to Wang et al. [15], Yamamoto et al. [16] and Duleba et al. [17], including the stability of the epigenetic programs underlying commitment, despite months of continuous in vitro propagation, the fact that these stem cells possess all of the information required to assemble into a higher order, and that regio-specific *3-D* epithelia are nearly identical to those found in the mature intestine in vivo. As stem cells comprise only a minor component of organoids, perhaps less than 1% [38], the molecular features of the stem cells of columnar epithelia, such as the intestinal tract, have remained unclear (Figure 2). Therefore, the selective cloning and proliferative expansion of highly clonogenic, ground-state intestinal stem cells through the Xian–McKeon method offer a first glimpse into the molecular properties of these cells. Our recent success in adapting this technology to clone stem cells in precancerous lesions, such as Barrett's esophagus, addressed the long-standing confusion in the cellular origin of Barrett's esophagus. Furthermore, using this technology, we generated broad sets of patient-matched stem cells corresponding to all stages of precursor lesions both in-line and out-line with the progression to adenocarcinoma. We can anticipate that each of these clones can be highly annotated with genomics, gene expression and differentiation fate information, accessible through a dynamic biorepository and thus be valuable substrates for new investigations across the cancer biology community.

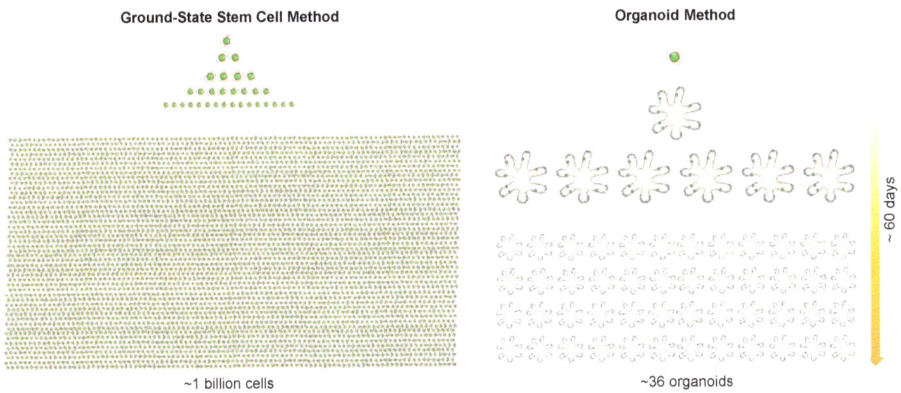

Figure 2. Rapid expansion of a single cell to one billion cells in approximately 60 days using the Xian–McKeon method. In comparison, one cell can become 36 organoids in the organoid method.

4. Biobanking of Ground-State Stem Cells and Personalized Regenerative Medicine

The onset of adult-stem-cell-based regenerative medicine started in the 1980s. Green and colleagues demonstrated the first example of cell therapy using cultured stem cells. They showed that human epidermis could be grown in the laboratory and transplanted to patients to reconstitute a functional epidermis [39,40]. Since then, transplantation of cultured epidermal stem cells has long been used to treat patients with burns, chronic wounds and stable vitiligo [41]. This is a life-saving procedure for patients with large area of burns. Moreover, the long-term effectiveness and safety of genetically-modified epidermal stem cells in correcting the severe skin blistering disease, epidermolysis bullosa, has been shown clinically [42].

In addition to skin, cultured stem cells from other epithelial tissues can be the source of stem-cell-based regenerative medicine. For example, a feature of lung regeneration that bodes well for regenerative medicine is that the underlying stem cell is highly clonogenic, which shows unlimited expansion capacity in vitro, and readily transplants to form functional alveoli in acutely damaged lungs [18]. We showed that a single p63+/Krt5+ DASC can be cloned, expanded and transplanted via intratracheal delivery to acutely damaged lungs, where they selectively inhabit damaged regions and differentiate to form Clara cells and alveoli composed of type I and type II pneumocytes. Importantly, these same p63+/Krt5+ DASCs showed no incorporation in mice without prior acute lung injury, suggesting that the efficient regenerative properties of these cells are not marred by "off-target" incorporation. Lastly, DASCs are readily cloneable from simple bronchoscopic biopsies, from bronchopulmonary lavage, or from transmural biopsies, providing good sources of autonomous stem cells that can be expanded to hundreds of billions of cells in weeks. Taken together, the established properties of DASCs, including clonogenicity, expandability, and facility for accurate transplantation obviate many theoretical objections that could have limited their use in regenerative medicine for either acute or chronic lung diseases.

It is also conceivable that the newfound ability to derive ground-state stem cells from columnar epithelium provides excellent sources for autologous transplantation to treat a wide range of disorders that current treatments are not able to help. For example, cultured ground-state intestinal stem cells may be of use to restore the intestinal epithelial functions following autologous transplantation in patients with severe forms of short bowel syndrome (SBS) [43], congenital disorders [44], or inflammatory bowel disease (IBD) [45,46].

In conclusion, we now have the technologies for cloning and culturing adult stem cells from nearly all types of epithelial tissues. Given the tremendous success of the pioneering work by Green and colleagues on the use of cultured adult stem cells in regenerative medicine, we should anticipate that more scientists and clinicians will recognize the potential of adult stem cells, appreciate the potential of biobanking various types of adult stem cells from individuals of diverse HLA haplotypes, and make attempts to use them for stem-cell-based personalized regenerative medicine.

Author Contributions: Conceptualization, W.X., M.V. and F.M.; Writing-Original Draft Preparation, W.X., M.D., and F.M; Writing-Review and Editing, W.X., Y.Y., M.V., M.D., and F.M.; Supervision, W.X. and F.M.; Funding Acquisition, W.X. and F.M.

Funding: This work was supported by grants from the Cancer Prevention Research Institute of Texas (CPRIT; RR15014 to WX and RR15088 to FM), the National Institutes of Health (1R01DK115445-01A1 to WX, and U24CA228550 to FM), the US Dept. of Defense (W81XWH-17-1-0123 to WX), the University of Texas Presidential Award (to WX) and the American Gastroenterology Association Research and Development Pilot Award in Technology (to WX).

Acknowledgments: We thank all the members in the Xian–McKeon laboratory for helpful discussions and support. We thank H. Green for advice and support.

Conflicts of Interest: W.X., F.M., M.D. and M.V. have filed a patent related to the technology used in the present work.

References

1. Müller, A.M.; Dzierzak, E.A. ES cells have only a limited lymphopoietic potential after adoptive transfer into mouse recipients. *Development* **1993**, *118*, 1343–1351. Available online: http://www.ncbi.nlm.nih.gov/pubmed/8269860 (accessed on 11 November 2018).
2. Helgason, C.D.; Sauvageau, G.; Lawrence, H.J.; Largman, C.; Humphries, R.K. Overexpression of HOXB4 enhances the hematopoietic potential of embryonic stem cells differentiated in vitro. *Blood* **1996**, *87*, 2740–2749. Available online: http://www.ncbi.nlm.nih.gov/pubmed/8639890 (accessed on 11 November 2018).
3. Bonde, S.; Dowden, A.M.; Chan, K.-M.; Tabayoyong, W.B.; Zavazava, N. HOXB4 But Not BMP4 Confers Self-Renewal Properties to ES-Derived Hematopoietic Progenitor Cells. *Transplantation* **2008**, *86*, 1803–1809. [CrossRef] [PubMed]
4. Iuchi, S.; Dabelsteen, S.; Easley, K.; Rheinwald, J.G.; Green, H. Immortalized keratinocyte lines derived from human embryonic stem cells. *Proc. Natl. Acad. Sci. USA* **2006**, *103*, 1792–1797. [CrossRef] [PubMed]
5. Amabile, G.; Welner, R.S.; Nombela-Arrieta, C.; D'Alise, A.M.; di Ruscio, A.; Ebralidze, A.K.; Kraytsberg, Y.; Ye, M.; Kocher, O.; Neuberg, D.S.; et al. In vivo generation of transplantable human hematopoietic cells from induced pluripotent stem cells. *Blood* **2013**, *121*, 1255–1264. [CrossRef] [PubMed]
6. Suzuki, N.; Yamazaki, S.; Yamaguchi, T.; Okabe, M.; Masaki, H.; Takaki, S.; Otsu, M.; Nakauchi, H. Generation of engraftable hematopoietic stem cells from induced pluripotent stem cells by way of teratoma formation., Molecular Therapy. *J. Am. Soc. Gene Ther.* **2013**, *21*, 1424–1431. [CrossRef] [PubMed]
7. Rheinwald, J.G.; Green, H. Serial cultivation of strains of human epidermal keratinocytes: The formation of keratinizing colonies from single cells. *Cell.* **1975**, *6*, 331–343. Available online: http://www.ncbi.nlm.nih.gov/pubmed/1052771 (accessed on 11 November 2018). [CrossRef]
8. Rama, P.; Matuska, S.; Paganoni, G.; Spinelli, A.; de Luca, M.; Pellegrini, G. Limbal Stem-Cell Therapy and Long-Term Corneal Regeneration. *N. Engl. J. Med.* **2010**, *363*, 147–155. [CrossRef] [PubMed]
9. Senoo, M.; Pinto, F.; Crum, C.P.; McKeon, F. p63 Is Essential for the Proliferative Potential of Stem Cells in Stratified Epithelia. *Cell* **2007**, *129*, 523–536. [CrossRef]
10. Kumar, P.A.; Hu, Y.; Yamamoto, Y.; Hoe, N.B.; Wei, T.S.; Mu, D.; Sun, Y.; Joo, L.S.; Dagher, R.; Zielonka, E.M.; et al. Distal Airway Stem Cells Yield Alveoli In Vitro and during Lung Regeneration following H1N1 Influenza Infection. *Cell* **2011**, *147*, 525–538. [CrossRef]
11. Matsuura, R.; Kogo, H.; Ogaeri, T.; Miwa, T.; Kuwahara, M.; Kanai, Y.; Nakagawa, T.; Kuroiwa, A.; Fujimoto, T.; Torihashi, S. Crucial Transcription Factors in Endoderm and Embryonic Gut Development Are Expressed in Gut-Like Structures from Mouse ES Cells. *Stem Cells* **2006**, *24*, 624–630. [CrossRef]
12. Sato, T.; Vries, R.G.; Snippert, H.J.; van de Wetering, M.; Barker, N.; Stange, D.E.; van Es, J.H.; Abo, A.; Kujala, P.; Peters, P.J.; et al. Single Lgr5 stem cells build crypt-villus structures in vitro without a mesenchymal niche. *Nature* **2009**, *459*, 262–265. [CrossRef]
13. Fordham, R.P.; Yui, S.; Hannan, N.R.F.; Soendergaard, C.; Madgwick, A.; Schweiger, P.J.; Nielsen, O.H.; Vallier, L.; Pedersen, R.A.; Nakamura, T.; et al. Transplantation of Expanded Fetal Intestinal Progenitors Contributes to Colon Regeneration after Injury. *Cell Stem Cell* **2013**, *13*, 734–744. [CrossRef] [PubMed]

14. Yin, X.; Farin, H.F.; van Es, J.H.; Clevers, H.; Langer, R.; Karp, J.M. Niche-independent high-purity cultures of Lgr5+ intestinal stem cells and their progeny. *Nat. Methods* **2014**, *11*, 106–112. [CrossRef] [PubMed]
15. Wang, X.; Yamamoto, Y.; Wilson, L.H.; Zhang, T.; Howitt, B.E.; Farrow, M.A.; Kern, F.; Ning, G.; Hong, Y.; Khor, C.C.; et al. Cloning and variation of ground state intestinal stem cells. *Nature* **2015**, *522*, 173–178. [CrossRef] [PubMed]
16. Yamamoto, Y.; Wang, X.; Bertrand, D.; Kern, F.; Zhang, T.; Duleba, M.; Srivastava, S.; Khor, C.C.; Hu, Y.; Wilson, L.H.; et al. Mutational spectrum of Barrett's stem cells suggests paths to initiation of a precancerous lesion. *Nat. Commun.* **2016**, *7*, 10380. [CrossRef] [PubMed]
17. Duleba, M.; Qi, Y.; Mahalingam, R.; Flynn, K.; Rinaldi, F.; Liew, A.-A.; Neupane, R.; Vincent, M.; Crum, C.P.; Ho, K.Y.; et al. An Efficient Method for Cloning Gastrointestinal Stem Cells from Patients via Endoscopic Biopsies. *Gastroenterology* **2018**. [CrossRef] [PubMed]
18. Zuo, W.; Zhang, T.; Wu, D.Z.; Guan, S.P.; Liew, A.-A.; Yamamoto, Y.; Wang, X.; Lim, S.J.; Vincent, M.; Lessard, M.; et al. p63(+)Krt5(+) distal airway stem cells are essential for lung regeneration. *Nature* **2015**, *517*, 616–620. [CrossRef] [PubMed]
19. Llames, S.; García-Pérez, E.; Meana, Á.; Larcher, F.; del Río, M. Feeder Layer Cell Actions and Applications. *Tissue Eng. Part B Rev.* **2015**, *21*, 345–353. [CrossRef]
20. Boyce, S.T.; Ham, R.G. Calcium-regulated differentiation of normal human epidermal keratinocytes in chemically defined clonal culture and serum-free serial culture. *J. Investig. Dermatol.* **1983**, *81*, 33s–40s. [CrossRef]
21. de Corte, P.; Verween, G.; Verbeken, G.; Rose, T.; Jennes, S.; de Coninck, A.; Roseeuw, D.; Vanderkelen, A.; Kets, E.; Haddow, D.; et al. Feeder layer- and animal product-free culture of neonatal foreskin keratinocytes: Improved performance, usability, quality and safety. *Cell Tissue Bank.* **2012**, *13*, 175–189. [CrossRef]
22. Lenihan, C.; Rogers, C.; Metcalfe, A.D.; Martin, Y.H. The effect of isolation and culture methods on epithelial stem cell populations and their progeny—Toward an improved cell expansion protocol for clinical application. *Cytotherapy* **2014**, *16*, 1750–1759. [CrossRef]
23. de Luca, M.; Pellegrini, G.; Green, H. Regeneration of squamous epithelia from stem cells of cultured grafts. *Regener. Med.* **2006**, *1*, 45–57. [CrossRef] [PubMed]
24. Green, H. The birth of therapy with cultured cells. *BioEssays* **2008**, *30*, 897–903. [CrossRef] [PubMed]
25. Kim, K.-A.; Kakitani, M.; Zhao, J.; Oshima, T.; Tang, T.; Binnerts, M.; Liu, Y.; Boyle, B.; Park, E.; Emtage, P.; et al. Mitogenic Influence of Human R-Spondin1 on the Intestinal Epithelium. *Science* **2005**, *309*, 1256–1259. [CrossRef] [PubMed]
26. Dreesen, O.; Brivanlou, A.H. Signaling pathways in cancer and embryonic stem cells. *Stem Cell Rev.* **2007**, *3*, 7–17. Available online: http://www.ncbi.nlm.nih.gov/pubmed/17873377 (accessed on 11 November 2018). [CrossRef] [PubMed]
27. Rock, J.R.; Onaitis, M.W.; Rawlins, E.L.; Lu, Y.; Clark, C.P.; Xue, Y.; Randell, S.H.; Hogan, B.L.M. Basal cells as stem cells of the mouse trachea and human airway epithelium. *Proc. Natl. Acad. Sci. USA* **2009**, *106*, 12771–12775. [CrossRef] [PubMed]
28. International Stem Cell Initiative; Amps, K.; Andrews, P.W.; Anyfantis, G.; Armstrong, L.; Avery, S.; Baharvand, H.; Baker, J.; Baker, D.; Munoz, M.B.; et al. Screening ethnically diverse human embryonic stem cells identifies a chromosome 20 minimal amplicon conferring growth advantage. *Nat. Biotechnol.* **2011**, *29*, 1132–1144. [CrossRef] [PubMed]
29. Avery, S.; Hirst, A.J.; Baker, D.; Lim, C.Y.; Alagaratnam, S.; Skotheim, R.I.; Lothe, R.A.; Pera, M.F.; Colman, A.; Robson, P.; et al. BCL-XL mediates the strong selective advantage of a 20q11.21 amplification commonly found in human embryonic stem cell cultures. *Stem Cell Rep.* **2013**, *1*, 379–386. [CrossRef]
30. Dickson, M.A.; Hahn, W.C.; Ino, Y.; Ronfard, V.; Wu, J.Y.; Weinberg, R.A.; Louis, D.N.; Li, F.P.; Rheinwald, J.G. Human keratinocytes that express hTERT and also bypass a p16(INK4a)-enforced mechanism that limits life span become immortal yet retain normal growth and differentiation characteristics. *Mol. Cell. Biol.* **2000**, *20*, 1436–1447. Available online: http://www.ncbi.nlm.nih.gov/pubmed/10648628 (accessed on 11 November 2018). [CrossRef]
31. Lee, G.; Hynes, R.; Kirschner, M. Temporal and spatial regulation of fibronectin in early Xenopus development. *Cell* **1984**, *36*, 729–740. Available online: http://www.ncbi.nlm.nih.gov/pubmed/6697394 (accessed on 11 November 2018).
32. Bissell, D.M.; Arenson, D.M.; Maher, J.J.; Roll, F.J. Support of cultured hepatocytes by a laminin-rich gel. Evidence for a functionally significant subendothelial matrix in normal rat liver. *J. Clin. Investig.* **1987**, *79*, 801–812. [CrossRef]

33. Barcellos-Hoff, M.H.; Aggeler, J.; Ram, T.G.; Bissell, M.J. Functional differentiation and alveolar morphogenesis of primary mammary cultures on reconstituted basement membrane. *Development* **1989**, *105*, 223–235.
34. Petersen, O.W.; Rønnov-Jessen, L.; Howlett, A.R.; Bissell, M.J. Interaction with basement membrane serves to rapidly distinguish growth and differentiation pattern of normal and malignant human breast epithelial cells. *Proc. Natl. Acad. Sci. USA* **1992**, *89*, 9064–9068. [CrossRef] [PubMed]
35. Broeckx, S.Y.; Maes, S.; Martinello, T.; Aerts, D.; Chiers, K.; Mariën, T.; Patruno, M.; Franco-Obregón, A.; Spaas, J.H. Equine Epidermis: A Source of Epithelial-Like Stem/Progenitor Cells with In Vitro and In Vivo Regenerative Capacities. *Stem Cells Dev.* **2014**, *23*, 1134–1148. [CrossRef] [PubMed]
36. Simian, M.; Bissell, M.J. Organoids: A historical perspective of thinking in three dimensions. *J. Cell Biol.* **2017**, *216*, 31–40. [CrossRef] [PubMed]
37. Ootani, A.; Li, X.; Sangiorgi, E.; Ho, Q.T.; Ueno, H.; Toda, S.; Sugihara, H.; Fujimoto, K.; Weissman, I.L.; Capecchi, M.R.; et al. Sustained in vitro intestinal epithelial culture within a Wnt-dependent stem cell niche. *Nat. Med.* **2009**, *15*, 701–706. [CrossRef] [PubMed]
38. Wang, F.; Scoville, D.; He, X.C.; Mahe, M.M.; Box, A.; Perry, J.M.; Smith, N.R.; Lei, N.Y.; Davies, P.S.; Fuller, M.K.; et al. Isolation and Characterization of Intestinal Stem Cells Based on Surface Marker Combinations and Colony-Formation Assay. *Gastroenterology* **2013**, *145*, 383–395. [CrossRef] [PubMed]
39. Green, H.; Kehinde, O.; Thomas, J. Growth of cultured human epidermal cells into multiple epithelia suitable for grafting. *Proc. Natl. Acad. Sci. USA* **1979**, *76*, 5665–5668. Available online: http://www.ncbi.nlm.nih.gov/pubmed/293669 (accessed on 11 November 2018). [CrossRef] [PubMed]
40. O'Connor, N.; Mulliken, J.; Banks-Schlegel, S.; Kehinde, O.; Green, H. Grafting of Burns with Cultured Epithelium Prepared from Autologous Epidermal Cells. *Lancet* **1981**, *317*, 75–78. [CrossRef]
41. Gallico, G.G.; O'Connor, N.E.; Compton, C.C.; Kehinde, O.; Green, H. Permanent Coverage of Large Burn Wounds with Autologous Cultured Human Epithelium. *N. Engl. J. Med.* **1984**, *311*, 448–451. [CrossRef]
42. Hirsch, T.; Rothoeft, T.; Teig, N.; Bauer, J.W.; Pellegrini, G.; de Rosa, L.; Scaglione, D.; Reichelt, J.; Klausegger, A.; Kneisz, D.; et al. Regeneration of the entire human epidermis using transgenic stem cells. *Nature* **2017**, *551*, 327–332. [CrossRef]
43. Buchman, A.L.; Scolapio, J.; Fryer, J. AGA technical review on short bowel syndrome and intestinal transplantation. *Gastroenterology* **2003**, *124*, 1111–1134. [CrossRef]
44. Hong, S.N.; Dunn, J.C.Y.; Stelzner, M.; Martín, M.G. Concise Review: The Potential Use of Intestinal Stem Cells to Treat Patients with Intestinal Failure. *Stem Cells Transl. Med.* **2017**, *6*, 666–676. [CrossRef] [PubMed]
45. Salim, S.Y.; Söderholm, J.D. Importance of disrupted intestinal barrier in inflammatory bowel diseases. *Inflamm. Bowel Dis.* **2011**, *17*, 362–381. [CrossRef] [PubMed]
46. Martini, E.; Krug, S.M.; Siegmund, B.; Neurath, M.F.; Becker, C. Mend Your Fences: The Epithelial Barrier and its Relationship With Mucosal Immunity in Inflammatory Bowel Disease. *Cell. Mol. Gastroenterol. Hepatol.* **2017**, *4*, 33–46. [CrossRef] [PubMed]

Journal of
Clinical Medicine

MDPI

Review

NANOmetric BIO-Banked MSC-Derived Exosome (NANOBIOME) as a Novel Approach to Regenerative Medicine

Bruna Codispoti [1], Massimo Marrelli [2], Francesco Paduano [1] and Marco Tatullo [1,*]

[1] Tecnologica Research Institute, 88900 Crotone, Italy; bruna.codispoti@tecnologicasrl.com (B.C.); francesco.paduano@tecnologicasrl.com (F.P.)
[2] Marrelli Health, 88900 Crotone, Italy; prof.marrelli@libero.it
* Correspondence: marco.tatullo@tecnologicasrl.com; Tel.: +39-349-874-2445

Received: 3 September 2018; Accepted: 12 October 2018; Published: 15 October 2018

Abstract: Mesenchymal stem cells (MSCs) are well known for their great potential in clinical applications. In fact, MSCs can differentiate into several cell lineages and show paracrine behavior by releasing endogenous factors that stimulate tissue repair and modulate local immune response. Each MSC type is affected by specific biobanking issues—technical issues as well as regulatory and ethical concerns—thus making it quite tricky to safely and commonly use MSC banking for swift regenerative applications. Extracellular vesicles (EVs) include a group of 150–1000 nm vesicles that are released by budding from the plasma membrane into biological fluids and/or in the culture medium from varied and heterogenic cell types. EVs consist of various vesicle types that are defined with different nomenclature such as exosomes, shedding vesicles, nanoparticles, microvesicles and apoptotic bodies. Ectosomes, micro- and nanoparticles generally refer to the direct release of single vesicles from the plasma membrane. While many studies describe exosomes as deriving from multivesicular bodies, solid evidence about the origin of EVs is often lacking. Extracellular vesicles represent an important portion of the cell secretome. Their numerous properties can be used for diagnostic, prognostic, and therapeutic uses, so EVs are considered to be innovative and smart theranostic tools. The aim of this review is to investigate the usefulness of exosomes as carriers of the whole information panel characterizing the use of MSCs in regenerative medicine. Our purpose is to make a step forward in the development of the NANOmetric BIO-banked MSC-derived Exosome (NANOBIOME).

Keywords: MSCs; EVs; exosomes; tissue regeneration; immunomodulation; biobanking

1. Introduction

Mesenchymal stem cells (MSCs) are almost ubiquitous in the adult body. Initially this population of immature cells has only been isolated in bone marrow, but further studies have described their presence in a wide range of organs (hearth, brain) tissues (fat, oral cavity), and liquid fluids (blood, urine, semen) [1,2].

MSCs are characterized by their elevated proliferation ability. Furthermore, their great plasticity confers them the potential to differentiate into mature specialized cells with a great variety of phenotypes (bone, cartilage, fat, neuron) [3,4], regardless of the original source of isolation.

The discovery that MSCs do not express HLA (human leukocyte antigen) class II has shed light on the low immunogenicity of these cells in allogeneic transplantation [5]. Further studies have illustrated that these cells are able to modulate the immunologic response and inflammation in an HLA-independent fashion [6].

Thus, the regenerative and immunomodulatory characteristics of MSCs have been exploited for the treatment of a multitude of clinical situations including the regeneration of traumatic injuries, autoimmune disease, cancer, neurologic disorders, and heart illnesses. More than 800 clinical trials have been approved for the most disparate clinical pathological conditions (clinicaltrials.gov) based on MSC therapy.

Although stem cell banks are widespread throughout the world, offering available samples to be transplanted as needed, MSC biobanking encounters many technical and regulatory issues linked to the risks associated with the use of cells in clinics.

The intense paracrine activity of MSCs has been strongly explored, and consists of cytokine and chemokine production, mitochondrial transfer, and extracellular vesicle secretion [7].

Extracellular vesicles are small double-lipid layer particles that are secreted by budding of the plasma membrane and released into the extracellular space. These nano- and micro-vesicles (such as exosomes, apoptotic bodies, nanoparticles, and microvesicles) carry proteins, mRNA, and miRNA among neighboring cells conveying a multitude of information, and stimuli. Many studies have focused on EVs investigation, mainly because these small structures embed the secretome of their originating cell. EVs secreted by MSCs have been studied in different model diseases in order to elucidate their great paracrine power for the regeneration of injured tissues and their significant immunomodulation ability.

EVs can be isolated from liquid fluids or harvested from a conditioned MSCs medium. Some investigations have tested the stability of EVs stored under different conditions as alternatives for cell biobanking. These should overcome some regulatory problems without altering the effectiveness of the induced biological functions [8].

Clinical trials are ongoing, with the aim of confirming the diagnostic and therapeutic potential of EVs. Many attempts must be made to shed light on the ideal stabilizer and cryoprotectant agent to be used and on the correct temperature degree to ensure effective long-lasting stability of stored EVs. The creation of a systematic process for EV biobanking could represent a very advantageous system for disposing of the powerful properties of MSC secretome, avoiding the typical issues linked to cell-therapy, and finally, moving towards the new frontiers of exosome-based therapy and diagnostics.

2. MSC Biobanking: Ethical, Technical, and Regulatory Issues

Biobanks for stem cell storage are growing in number in order to satisfy the increasing needs for available immature populations for research and clinical use.

Embryonic stem cells, in the early blastocyst stage, are able to differentiate in all organs and tissues of the body (totipotency); fetal stem cells can differentiate into tissues of different germ layers (pluripotency). Despite the important potential for clinical application, many ethical concerns limit the use of these cells from human sources [9]. Moreover, another limit of these highly immature populations is the real risk of insurgence of teratomas [10].

Likewise, stem cells residing in the adult body have a differentiation potential limited to a relatively finite number of specialized tissues (multipotency). Their isolation and application bypass the ethical matters linked with embryonic and fetal stem cell sources [11].

Adult stem cells are unable to give rise to teratomas, but some evidence indicates the advent of secondary solid tumors after hematopoietic stem cell transplantations from bone marrow [12,13]. The United States of America agency—Food and Drugs Administration (FDA) established basal concerns for stem cell manipulation, including the definition of sample identification, in order to evaluate the autologous or unrelated setting. The absence of extraneous substances is potentially capable of altering cell behavior and efficacy. The absence of microorganism contamination is necessary to avoid disease transmission. Furthermore, the sample should be successfully efficient to promote the required biological effects [14].

In addition to this, many issues arise from the definition of stem cell use as being a common "practice of medicine" or as a "biological drug"; in the latter case, procedures must be suitable for FDA approval.

Despite some clinicians asserting that stem cell products are not drugs and that they should not be regulated by the Food and Drug Administration, FDA regulations argue that these clinics generally produce stem cell products that can be classified as biological drugs. The approval for production and clinical use of stem cell biological drugs implies acceptance by the FDA of an Investigational New Drug (IND) application that requires a Biological License Application (BLA) (http://www.ipscell.com/2015/01/stemhumanexperiment).

Furthermore, isolation of MSCs from different sources often does not allow an adequate clinical number of cells to be directly transplanted [15]; hence, the passage of in vitro expansion is frequently required. This handling step is in contrast with the FDA key concept of "minimal manipulation" that distinguishes clinical products from biological drugs that, by definition, require "more than minimal manipulation" processes [16]. Another central concept for the FDA is the "homologous use", that is, the application of stem cells derived from a specific tissue source, to transplantation in the same tissue of isolation, for example, blood-derived MSCs can only be used for hematologic applications (http://www.ipscell.com/2015/01/stemhumanexperiment).

The most commonly employed sources of MSCs for biobanking are adipose tissue and umbilical cord tissues, probably due to the easy accessibility and reduced invasiveness required for cell collection. Each cell source requires specific protocols and guidance approval for safe and high quality isolation, manipulation, cryopreservation and storage [14]. During the medical banking of MSC, particular consideration should be given to the process of cryopreservation, for example, the reproduction of perfect temperature variation during freeze and thaw passages, durable storage in pharmaceutical-grade liquid nitrogen, the optimal selection of a cell storage container, the freezing device used and the adequate and approved composition of cryopreservation media [17].

3. Extracellular Vesicles

Extracellular vesicles (EVs) are secreted by virtually all cell types and have been found in vivo in biologic fluids [18] including blood, urine, saliva, breast milk, ascites, cerebrospinal fluid, and semen [19]. EVs from the cell membrane are present in a wide size range; they can be large (1 μm), a few micrometers long, or up to about 150 nanometers long [20]. Based on their dimensions, EVs are defined as microvesicles (MVs) or nanovesicles. Other authors use different definitions such as micro and nanoparticles, shedding vesicles, apoptotic bodies, exosomes, endosomes or names derived from the cell of origin, such as oncosomes, which are secreted by tumor cells. The mechanism of EV formation includes endosomal formation and maturation by internal budding in endosomal space of multiple vesicles called intraluminal vesicle (ILV). The late-formed endosomes turn into a multivesicular body (MVB) that fuses with the plasma membrane (PM) and releases their internal vesicle content as exosomes by a mechanism of exocytosis [21]. Alternatively, EVs could be directly released to the extracellular space by outward budding from the plasma membrane surface [22] (Figure 1). Many authors define exosomes as the EVs secreted from MVBs while single vesicle budding refers to microvesicles. These definitions are yet to be collectively accepted.

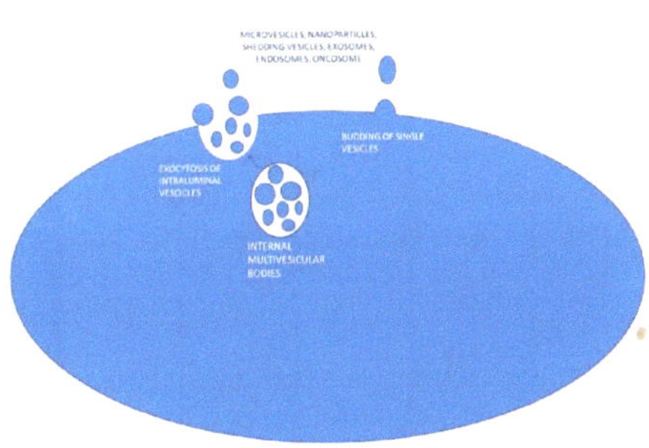

Figure 1. Proposed mechanisms for extracellular vesicle secretion.

3.1. Exosomes: The Good, the Bad and the Ugly

In the 1980s, exosomes were described as vesicles of endosomal origin with sizes ranging from 30 to 1000 nm. In 1987, the term "exosome" was first used to describe small membrane vesicles formed by the vesiculation of intracellular endosomes and released by exocytosis [23].

These vesicles are released during reticulocyte differentiation as a consequence of the fusion of multivesicular bodies (MVBs) with the plasma membrane [24,25]. At the time of their discovery, exosomes were thought to be "storage bodies" for the cell's unwanted components. In 1996, Raposo et al. demonstrated, by electron microscopy (EM), the presence of secreted exosomes containing MHC class II molecules, previously labeled by BSA gold tracer, in Epstein–Barr virus (EBV)-transformed B cell lines. Furthermore, the protein composition of the secreted exosomes was shown to differ from that of the plasma membrane. These findings suggested the potential endocytic origin of exosomes and led to new interesting hypotheses on the exosome function [26].

In the 1990s, various studies on blood cells proposed the involvement of the endosome in central cellular processes, such as blood coagulation, intercellular communication, lipid metabolism [27,28], in the release of inflammatory mediators [29] and in the proliferation of endothelial and immunologic cells [30].

The discovery of tumor-derived extracellular vesicles attracted scientific interest in the involvement of extracellular vesicles in suppressing the immune system against a tumor [31,32], but also in activating the immunologic anti-tumor response [33]. Moreover, tumor exosomes have been shown to participate in cancer angiogenesis [34] and in the spreading of metastatic cells [35].

These interesting findings prompted those studying EVs to establish a database named *Vesiclepedia* that includes data on mammalian exosomes (http://microvesicles.org) [36] and also another database including studies of non-mammalian EVs (http://evpedia.info) [37]. Both databases are continuously updated, making them an important instrument to improve EV knowledge.

3.2. Exosomes: Individuation and Isolation Strategies

The initial attempts to purify exosomes basically consisted of differential centrifugation, in which successive centrifugation steps are performed at increasing speeds, allowing the recovery of fractions with decreasing dimensions, cells, dead cells, and cellular debris. Then, a final ultracentrifugation at 100,000× *g* allows exosome recovery, followed by washing for the removal of protein aggregates [38]. Alternative protocols have been used to replace the differential centrifugation steps with a single filtration step, such as the use of 0.22 μm filters, an increase in the ultracentrifugation speed to 140,000× *g* [39], or size exclusion chromatography to recuperate units larger than 50,000 kDa, permitting the segregation of soluble proteins [40]. The inclusion of an extra purification step using a sucrose gradient has been used to determine the sedimentation of protein aggregates through sucrose, while vesicles float into a specific position within the sucrose gradient. These methods allow the aggregates of proteins to be separated from membrane-enclosed vesicles that could be available for therapeutic use [41]. Recently, many commercially available kits have been produced by companies for EV isolation that ensure quick and easy purification protocols exploiting immune labeling with magnetic beads, the use of specific filters, and polymer-based precipitation. These procedures, moreover, allow the direct recovery of protein and/or nucleic acids carried by exosomes. The choice of the best method for EV isolation is related to the source of exosomes (i.e., biological fluid specimen, cell supernatant) and the type of analysis to be performed, such as observation, enumeration, flow cytometer investigation, proteomic studies, or RNA isolation. Due to the 200 nm resolution limit of classical optical microscopes, EVs are mainly observed by electron microscopy (EM)—the election technique to observe the small sizes and morphologies of exosomes [26]. Nanoparticle tracking analysis (NTA) is a device that is capable of statistically calculating the diameter of laser-illuminated individual particles by tracking their movement under Brownian motion [42], allowing the evaluation of the size and distribution of EVs. Fluorescent labeling of vesicles using lipid dyes allows the identification of aggregates and large size EVs by fluorescent microscopy [43]. The last generation flow cytometers are able to identify microparticles in the forward scatter channel, but with scarce discrimination efficacy [44].

3.3. Exosomes: Composition and Contents

Exosomes are composed of an external lipid bilayer that is mainly enriched in saturated fatty acids, sphingomyelin, phosphatidylserine, cholesterol, and ceramides [45]. Interestingly, the exposition of phosphatidylserine on the surface membrane of exosomes which, in live cells, is confined to the inner leaflet of the PM, has been used to characterize these membrane vesicles by binding with annexin V [46].

Another component of exosomes is cytoplasmic and transmembrane proteins. The protein pattern is often the same as that of the deriving cell. Specifically, this is the case for proteins from the cytosol and the plasma membrane; however, peptides typical of other cellular organelles are absent, thus sustaining the endosomal origin of EVs. Some specific protein subsets are shared by exosomes derived from heterogeneous cell types [47]. In 2007, Valadi et al. detected the presence of mouse proteins in human cells fed with mouse exosomes. These peptides were shown to be absent in exosomes; therefore, it has been hypothesized that exosomes carry the related genetic information. These results demonstrate that exosomes also carry mRNA that is able to be translated into proteins [48]. In 2008, miRNAs were identified in mixed EVs derived from glioblastoma and blood cells [49,50]. Recent NSG techniques elucidated the presence of ulterior genetic material embedded in EVs, including noncoding RNA, with potential regulatory effects on the genomes of target cells [51]. Recent research has demonstrated changes in the composition of EVs after environment alteration mimicking pathological conditions. These changes concern alterations in the protein and RNA content [52,53], as well as changes in the lipid composition [54] of the EV membrane.

Despite the numerous model disease-based experimental works that have demonstrated that the functional effects of MSC exosomes are carried by their RNA content, other works have suggested that MSC exosomes most probably work through the protein rather than the RNA, due to an inadequate RNA configuration and/or concentration. However, for this to be true, proteins would need to be present in MSC exosomes in a representative therapeutic dose capable of producing a biologically relevant response, especially for the catalytic activity of enzymes [55]. Further investigation is needed to better understand the roles and related effects of the different components of EVs.

3.4. Exosomes: Biogenesis and Secretion

The formation of exosomes follows an endocytic process that consists of the internalization of extracellular elements. The formed early endosome starts a maturation process consisting of the internal budding of intra lumen vesicles. These late endosomes are referred to as multi-vesicular bodies. The MVBs could be addressed through lysosomal degradation or, to a lesser extent, by exocytosis [56]. A possible mechanism of intra lumen vesicle formation involves the endosomal sorting complex required for transport (ESCRT), and a multiprotein machinery organized into four subunits (ESCRT 0, 1, 2 and 3) that is associated with supportive proteins (VPS4, VTA1, ALIX also called PDCD6IP) [57]. Interestingly, the inactivation of the entire ESCRT complex does not limit MBV formation; thus, further mechanisms should be concomitantly active [58]. These alternative methods may involve the tetraspanin CD63 [59], or the lipid metabolism enzymes neutral sphingomyelinase [60], and phospholipase D2 [61].

Multiple mechanisms are involved in EV secretion. For example, elevation of the intracellular calcium levels has been documented to induce secretion [62]. Tumor cells spontaneously secrete exosomes (oncosomes) with invasive properties [43], but changes in extracellular conditions could also trigger EV secretion. The interaction between dendritic cells and $CD4^+$ T lymphocytes has been demonstrated to induce exosome secretion [63]. The release of neurotransmitters stimulates oligodendroglial exosome secretion that is internalized by neurons through endocytosis [64].

The budding of intraluminal vesicles implies interactions with the cytoskeleton and binding with PM or other membranous compartments that are mediated by a family of GTPases named Rab proteins [65]. This is followed by a SNARE–SNAPs protein interaction [66] that mediates membrane fusion and exosome release. Many other molecules are involved in biogenesis and in intra- and extracellular EV secretion, and the related cellular processes remain a matter of debate.

The limited dimensions of EVs makes their tracking difficult after budding in the extracellular space. To exert their functions on target cells, exosomes must be captured and internalized. The specific binding could be mediated by ligands and receptors present in both EVs and recipient cells. The blocking of the integrins tetraspondin, ICAM-1 and LFA-1 with specific antibodies partially inhibits the interaction of dendritic cell-derived EVs with recipient dendritic cells [67], and heparan sulphate proteoglycans expressed by tumor-EVs are recognized by tumor cells [68]. The final step is the transfer of EV content into the recipient cell. This passage could be done through either an endocytic or phagocytic method, followed by the degradation of; the internalized vesicles to extract their components. The contents of exosomes could also be directly released into the cytosol. This is the case for nucleic acids that influence the gene expression of target cells, for instance, the fusion of EVs with the plasma membrane or with the endocytic membrane must occur. Abrami et al. demonstrated that after endocytosis of the anthrax lethal toxin, these components were found in the intraluminal space of MVBs and in the cytosol of target cells. Moreover, they could be delivered to the extracellular medium as exosomes and infect neighboring cells. These results functionally demonstrate membrane fusion [69,70].

4. Exosomes from MSCs

The secretory ability of MSCs has been well established by enumerable scientific reports; thus, the identification of exosomes as carriers of stimuli and information among cells suggests a central role of EVs as effectors of the paracrine activity of MSCs (Table 1).

As previously explained, exosomes mainly exert the same features as the cells they came from, such as the transport of proteins, miRNA, mRNA, and other soluble factors implied by MSCs' functions, including immunomodulation and tissue regeneration [70].

Early evidence describing the implication of exosomes in MSCs' paracrine activity was illustrated in a model of acute kidney injury. The authors demonstrated that microvesicles derived from human bone marrow MSCs stimulate proliferation and induce resistance to apoptosis in tubular epithelial cells. The administration of MSC microvesicles into SCID (severe combined immunodeficiency) mice accelerates functional recovery after glycerol-induced acute kidney injury. Moreover, the administration of RNAse abolishes the described in vitro and in vivo regenerative effects. These authors concluded that the RNA content of the administrated MSC microvesicles activates proliferation and injury recovery [71].

It is important to note that RNAse acts only in conjunction with detergent to lyse the cholesterol-rich phospholipid membrane of vesicles. This shell, in fact, guarantees the protection of exosome content from extracellular insults during passages among cells.

Another study described the exosome-induced recovery effects in a mouse model of myocardial ischemia/reperfusion injury. In this work, EVs of 50- to 100-nm sized particles secreted from MSCs were visualized by electron microscopy and purified by size exclusion fractionation on a HPLC. The administration of the isolated exosomes was shown to reduce the infarcted area in mice [72]. The properties of MSC-derived exosomes have been further explored in several disease models; some investigations focused on the composition and morphology of MSC exosomes.

Exosomes secreted from MSCs express typical mesenchymal markers such as CD105, CD29, CD90, CD44, and CD73, together with more exosome-specific surface antigens including CD107, CD63, CD9, and CD81 [73].

The proteomic profile studies performed by using mass spectrometry and antibody array revealed the different protein compositions in different HPLC-purified MSC-derived exosome preparations with only 20% correspondence in all three batches. The same study identified the presence of a proteasome subunit in MSC exosomes. These findings not only suggest proteasome as a direct actor in contrasting diseases, but also indicate that the proteasome activity could explain the changes in exosome composition during a cell's life span [74]. Other studies investigated the miRNA contents of MSC-derived exosomes. Tian Sheng Chen et al. suggested that MSCs could enable miRNA-mediated communication among cells by secreting microvesicles rich in miRNA. Microarray analysis and Q-RT PCR results showed that these miRNAs encapsulated in MSC-derived microvesicles were predominantly present as pre-miRNA and thus, in their precursor form [75].

4.1. Exosomes from MSCs: Regenerative Potential

The regenerative potential of exosomes derived from MSCs has been investigated in various model diseases and in many different works referred to kidney, liver, heart, and neural injuries.

EVs secreted from mesenchymal stem cells carry several active molecules that have been implicated in central cell regenerative mechanisms including the prevention of cell apoptosis, the promotion of cell proliferation, and the improvement of neovascularization [76,77].

In a study of neurotoxicity induced by smoke and HIV, the neuroprotective potential of exosomes derived from astrocytes carrying antioxidants was proposed to protect neurons against oxidative damage [78]. Another study revealed the presence of neural growth factor transcripts in exosomes from adipose derived-MSCs through verification by immunohistochemical techniques and footprint analysis. These EVs were able to promote neurite outgrowth in vitro and increase regeneration in vivo after sciatic nerve injury [79]. Research on Alzheimer's disease has shown that exosomes secreted by NCS have a promoting role in beta-amyloid production and clearance; hence, exosomes have been proposed as a target in Alzheimer's therapy [80].

In an interesting work which challenged the notion that ROS have an exclusively nerve degenerative function, axonal regeneration and functional recovery after spinal injury were demonstrated to be induced by NOX2 (nitric oxide) carried by macrophage-secreted exosomes. Thus, NOX2 may induce the oxidation and inactivation of PTEN (Phosphatase and tensin homolog), leading to PI3K (Phosphatidyl-Inositol 3-Kinase)-phosphorylated (p-) protein kinase *B* signaling and activation of regenerative progress [81].

MSC exosomes have been shown to promote skeletal muscle regeneration by inducing miRNA-mediated myogenesis and angiogenesis in vitro and muscle regeneration in a mouse model of muscle injury [82].

Mesenchymal stromal cell-derived fractionated secretomes enriched in exosomes enhance recovery in in-vitro and in vivo liver injury models [83].

A resident population of MSCs found within the renal glomeruli was shown to contribute to the stimulation of ischemia-reperfusion-related acute kidney injury (AKI) recovery in SCID mice through the release of EVs [84]. RNAse treatment decreased the described regenerative effects. A down-regulation of miRNA production induced by Drosha knockdown contracted the intrinsic regenerative potential of mesenchymal stromal cell-derived EVs in a mouse model of glycerol-induced acute kidney injury [85]. This evidence suggests the central role of RNA in MSC-EV-mediated AKI recovery.

Intramyocardial injection of EVs released by MSCs enhanced the in vitro proliferation, migration, and tube formation of endothelial cells, improved blood flow recovery, and reduced infarct size in an acute myocardial infarction rat model [76]. Exosomes isolated from cord blood-derived, Akt-overexpressing MSCs promoted angiogenesis and lesion size reduction in a rat model of acute myocardial infarction by angiogenesis activation mediated by *PDGF-D* that was up-regulated in the Akt exosomes [86]. Exosomes derived from MSCs overexpressing *GATA-4* improved survival and reduced apoptosis in a culture of rat neonatal cardiomyocytes, as well as restoring cardiac contractile function, and reducing infarct size in vivo. These effects have been attributed to the enhanced expression of miR-19a that down-regulates PTEN, thereby activating the Akt and ERK signaling pathways [87].

Recently, researchers have focused their studies on human embryonic stem cell-derived MSCs (ESC-MSCs). Specifically, the components contained in their conditioned medium were carefully analyzed by multidimensional protein identification technology, cytokine antibody array, gene microarray, and quantitative RT-PCR assays. Computational analysis of the obtained data predicted the presence of gene products involved in metabolism, the immunological response, and in tissue differentiation, including neo-angiogenesis, hematopoiesis, and bone formation [88]. As previously described, exosomes derived from ESC-MSCs showed the ability to restore cardiac functionality after myocardial ischemia/reperfusion injury and severe infarction [72,74]. Furthermore, exosomes were reported to have a key role in promoting miRNA-mediated cell-to-cell communication [75].

Despite the clear demonstration of the effects of MSC-derived exosomes in several clinical conditions, the induction of pluripotent stem (iPS) cell-derived exosomes represents a novel field of interest.

A research team investigated exosomes obtained from human embryonic stem cell-induced mesenchymal stem cells (ESC-MSCs) and used them in a mouse model affected by the destabilization of the medial meniscus. Their results showed a significant reduction of the induced osteoarthritis mediated by ESC-MSCs, exerted through regulation of the cartilage synthesis/degradation process [89].

Zhang et al. proposed a 'cell-free' therapeutic approach for the treatment of osteochondral defects, demonstrating the ability of human embryonic MSC-derived exosomes to repair cartilage and subchondral bone injuries [90]. An interesting paper by Lai RC et al. proposed the large-scale production of exosomes from human ESC-MSCs after their C-Myc-mediated immortalization, challenging the conventional concept of pharmaceutical manufacturing and introducing the role of MSCs as "producers" of therapeutics [91].

Some authors investigated the regenerative potential of exosomes secreted by iPSC-derived MSCs. Such exosomes were shown to promote neovascularization after excision of the femoral artery in a mouse model of hind-limb ischemia, thus demonstrating a protective effect on limb ischemic injury [92].

In a rat model of a skin wound, exosomes released by human-induced pluripotent stem cell-derived MSCs were shown to be capable of stimulating cutaneous wound healing through the promotion of collagen production and by triggering angiogenesis [93]. The intravenous injection of iPS-MSC-derived exosomes into a steroid-induced osteonecrosis of the femoral head rat model importantly limited bone degradation. Finally, in vitro experiments on iPS-MSC-derived exosomes showed an increase in the migration of endothelial cells and proliferation mediated by PI3K/Akt signaling pathway activation [94].

4.2. Exosomes from MSCs: Immunomodulation

The ability to regulate the immune system is a central feature of MSCs and is mainly exerted by their important paracrine activity which involves the release and transport of bioactive molecules that could be mediated by extracellular vesicle secretion.

Exosomes secreted by MSCs exert direct effects on immune cells. The addition of MSC-EVs in a culture of peripheral blood mononucleated cells reduced the growth and differentiation of B cells and decreased the production of IgM, IgG, and IgA [95]. In a culture of splenic mononuclear cells from a mouse model of autoimmune encephalomyelitis, exosomes from MSCs were shown to counteract auto-reactive lymphocyte expansion, induce apoptosis of activated T cells, and initiate Treg production and the secretion of IL-10 and TGF-β anti-inflammatory cytokines [96]. Additionally, in monocytes/macrophages, the administration of MSCs-EVs was shown to increase the levels of IL-10 anti-inflammatory cytokines and reduce the secretion of pro-inflammatory interleukins 1β, 6, 12p40, and TNF-α [97]. Furthermore, different works have described the induction of the M1 to M2 phenotypic transition of macrophages treated with EVs [98,99].

The immunomodulatory effects of MSC-EVs have been investigated in various disease models. Exosomes secreted by MSCs up-regulate IL-10 and TGF-β1 in peripheral blood mononuclear cells isolated from asthmatic patients, thus stimulating the proliferation and immune-suppression capacity of regulatory T lymphocytes. These effects led to the alleviation of inflammation in asthma disease [100].

Exosomes derived from human umbilical cord MSCs were shown to restore organ function in a model of carbon tetrachloride (CCl_4)-induced fibrotic liver by reducing collagen deposition and decreasing (TGF)-β1 and Smad2 expression, leading to an inhibition of epithelial-to-mesenchymal transition, thus opposing inflammation and fibrosis [101].

In a mouse model of renal ischemia/reperfusion injury, macrophage activation was suppressed by exosomes from MSCs that express high levels of CLR2 (Cryptic loci regulator 2). CLR2-expressing exosomes were shown to be capable of binding free CCl_2 and abolishing their activity in macrophage recruitment, thereby down-regulating inflammation [102]. Liu et al. demonstrated that exosomes from bone marrow-derived MSCs possess strong pro-angiogenic properties, reduce neuronal cell apoptosis, stimulate axonal regeneration, suppress inflammation, and attenuate lesion size after traumatic spinal cord injury (SCI) [103]. Exosomes derived from human cord blood derived-MSCs increase the functional recovery after SCI through the down-regulation of inflammatory cytokines, including IL-6, TNF-α, IFN-γ, and MIP-1α [104].

An interesting report highlighted the role of a particular class of EVs derived from apoptotic bodies, which they call ApoEVs, that is able to modulate the immune system in both activating and suppressing ways [105]. ApoEVs strongly interact with antigen-presenting cells by direct or cross-presentation mechanisms and promote the clearance of apoptotic cells by recruiting phagocytic cells. Related effects may include the production of autoantibodies, consequently favoring autoimmune conditions such as systemic lupus erythematous [106] although the reduced clearance of dying cells has been described as a promoter of autoimmunity [107]. Furthermore, ApoEVs could stimulate antitumor immunity, with exosomes from melanoma cells facilitating the passage of tumor antigens to antigen-presenting cells [108]. The same mechanisms provide a protective effect in infective conditions. ApoEVs secreted by tuberculosis-infected macrophages carry microbial-derived antigens to antigen presentation cells that recruit CD4 and CD8 T lymphocytes [109].

Table 1. Mesenchymal stem cells (MSC)-derived extracellular vesicles (EVs).

EVs Definition	Tissue of Origin/Exosome Source	Content/Composition	Functions	References
Microvesicles	Human bone marrow MSCs	mRNA	Protection against acute tubular injury	[71]
Exosomes	Human ESC-derived mesenchymal stem cell		Decreases myocardial ischemia/reperfusion injury	[72]
Exosomes	Human ESC-derived mesenchymal stem cells	20S proteasome	Reduce the accumulation of misfolded proteins in a mouse model of myocardial infarction	[74]
Microparticles	Human embryonic stem cell (hESC)-derived MSC	pre-miRNA	Enable miRNA-mediated intercellular communication	[75]
Extracellular vesicles	Human bone marrow mesenchymal stem cells		Stimulate neoangiogenesis, preserve cardiac function in a rat myocardial infarction model	[76]
Exosomes	Human umbilical cord mesenchymal stem cells		Improves recovery in cisplatin-induced acute kidney injury rat model	[77]
Exosomes	Rat Adipose-Derived Mesenchymal Stem Cells	Neural growth factors transcripts	Increase neurite outgrowth in vitro and enhance regeneration after sciatic nerve injury in vivo	[79]
Exosomes	Macrophages	Functional NADPH oxidase 2 complexes	Axonal regeneration through a NOX2-PI3K-p-Akt signalling pathway	[81]
Exosomes	Human bone-marrow-derived mesenchymal stem cells	Repair-related miRNAs, miR-494	Stimulation of myogenesis and angiogenesis in vitro, and muscle regeneration in an in vivo model of muscle injury	[82]

Table 1. *Cont.*

EVs Definition	Tissue of Origin/Exosome Source	Content/Composition	Functions	References
Exosomes	Rat bone marrow MSCs	Exosome enriched mesenchymal stromal cell-derived fractionated secretome	Repair and healing of injured liver tissue	[83]
Extracellular vesicles	MSCs within the glomeruli (Gl-MSCs)	miRNAs	Recovery of in a mouse model	[84]
Extracellular vesicles	Bone Marrow MSCs	microRNA	microRNA depletion in EVs from MSCs decrease their intrinsic regenerative potential in acute kidney injury	[85]
Exosomes	Akt-Modified Human Umbilical Cord Mesenchymal Stem Cells	Platelet-derived growth factor D (PDGF-D)	Improve angiogenesis and promote cardiac regeneration	[86]
Exosomes	GATA-4 overexpressing mesenchymal stem cells	Anti-apoptotic microRNAs, miR-19a	Cardio-protection	[87]
Exosomes	Human embryonic stem cell-induced mesenchymal stem cells		Relieve osteoarthritis through the regulation of synthesis/degradation of cartilage extracellular matrix	[89]
Extracellular vesicles/exosomes	Human embryonic mesenchymal stem cells		Stimulation of osteochondral regeneration and repair	[90]
Exosomes	Human induced pluripotent stem cells-derived MSCs		Alleviate hind-limb ischemia and stimulate angiogenesis in mice	[92]
Exosomes	Human induced pluripotent stem cells-derived MSCs		Stimulation of angiogenesis and collagen synthesis accelerating cutaneous wound healing in rats	[93]
Exosomes	Human induced pluripotent stem cells-derived MSCs		Increasing in endothelial cells migration and proliferation and reduction of bone degradation in an osteonecrosis of the femoral head rat model	[94]
Membrane vesicles	Bone marrow human MSCs		Immunosuppressive effect on B lymphocytes	[95]
Microvesicles	Murine bone-marrow derived MSCs	PD-L1, galecin-1 and membrane-bound TGF-β	Initiation of peripheral tolerance and regulation of immune responses	[96]
Exosomes	Human embryonic stem cell (ESC)-derived MSCs		Prolonging of the survival of allogenic skin graft in mice and increased Tregs.	[97]
Extracellular vesicles	Human adipose derived-MSCs		Exert anti-Inflammatory effects, stimulate macrophages switching from a M1 to a M2 phenotype	[98]
Extracellular vesicles	Porcine adipose tissue-derived MSCs	Anti-inflammatory cytokine interleukin (IL) 10	Reduction of renal inflammation, increasing of medullary oxygenation in porcine model of metabolic syndrome and renal artery stenosis	[99]
Exosomes	Human bone-marrow derived mesenchymal stem cells		Promotion of upregulation of IL-10 and TGF-β1 from PBMCs, stimulation of proliferation and immune-suppression capacity of Tregs in asthmatic patient	[100]
Exosomes	Human umbilical cord mesenchymal stem cells		Amelioration of liver function and restoration of liver fibrosis	[101]
CCR2 positive exosomes	Bone marrow mouse mesenchymal stem cells	C-C motif chemokine receptor-2 (CCR2)	Establishment of protective effects on renal ischemia/reperfusion injury in mouse	[102]
Exosomes	Bone marrow mesenchymal stem cells from rats		Attenuation of inflammation, glial scar formation and of neuronal cells apoptosis, lesion size reduction, improving of axonal regeneration, and of functional recovery after traumatic Spinal Cord Injury in rat model	[103]
Exosomes	Human umbilical cord mesenchymal stem cells		Stimulation of spinal cord injury healing via mitigating the inflammation at the injured area	[104]

5. Biobanking of Exosomes

As described above, exosomes are acquiring even more importance in the clinic due to their central roles as diagnostic and prognostic biomarkers [110], therapeutic targets [111], and drug vehicles [112].

The possibility of storing exosomes in biobanks could represent an advantageous system for various medical applications as well as for research purposes (Figure 2).

"NANOBIOME" is the acronym of NANOmetric BIO-banked MSCs-derived Exosome, a novel approach that is based on the biobanking of exosomes secreted by MSCs as opposed to managing batches of MSCs. Furthermore, the biological role of exosomes as the holder of the secretome and as the external carrier of the functional biological effects of their generating cells confers to exosomes a significant potential role for innovative "cell-free" regenerative medicine.

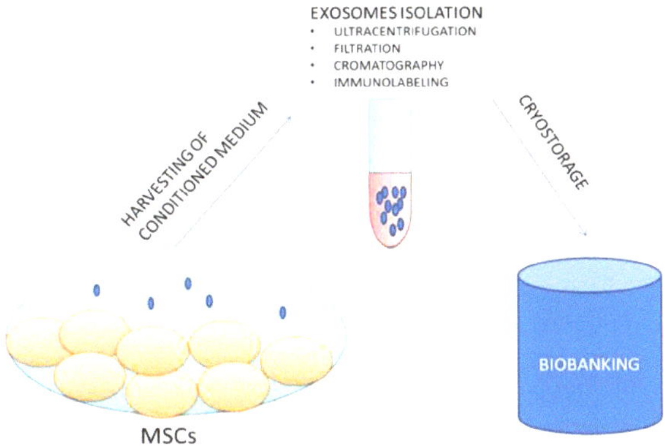

Figure 2. The proposed NANOBIOME approach.

Biobanks can currently be defined as facilities where high-quality biological samples are acquired, processed, and preserved by long-term storage for clinical distribution or for future research investigations [113].

Each sample collected and stored in a biobank has to be inserted in a database containing precise information about demographic and clinical data [114].

Liquid fluids (peripheral blood, serum, plasma, urine, saliva, semen), solid tissues, cells (isolated peripheral blood mononuclear cells (PBMCs), stem cells, and other cell types), and biomolecules (RNA, and DNA) are only some of the variety of sample types that are processed and stored in biomedical and research biobanks.

The increasing significance of exosome use in cancer, metabolic diseases, traumatic injuries, and other complex diseases has made this biocomponent amenable for biobanking, and it has great potential advantages for the diagnosis and treatment of these particular diseases [115].

Exosomes exert the ability to cross biological barriers and to selectively reach target cells and organs due to their intrinsic homing ability. Moreover, they are well tolerated by the human body with proper circulation times, and finally, their membranes can be modified to express or incorporate specific molecules. All of these benefits could be exploited for drug delivery [116].

The biobanking process requires the formulation of defined protocols for the collection, sampling, and storage of samples.

The reduced size of exosomes limits their morphologic observation with conventional optical microscopes; thus, their identification has been mainly assessed by flow cytometry, electron microscopy, and NTA methods. Furthermore, EV harvesting may require different centrifugation techniques associated or not with the filtration and chromatography steps. Immunolabeling is instead exploited by numerous commercially available kits specific for the isolation of exosomes from different biological sources.

Despite much scientific evidence reporting procedures for EV identification and isolation, less information concerning the methods needed for cryopreservation and the storage stability of exosomes is available. Different procedures are required depending on the source of isolation [8].

Kalra et al. tested three different methods for the isolation of exosomes from human plasma, with better results being obtained for the OptiPrepTM density gradient method. Furthermore, they investigated the stability of harvested exosomes for 90 days at different storage temperatures, with superior results being shown for cryopreservation at −80 °C [117]. The stability of exosomes derived from HEK 293T, ECFC, and MSCs cell types was observed during storage at 37 °C, 4 °C, and −20 °C. The initial particle size of 110 nm was assessed by scanning SEM and dynamic light scattering NTA analyses. Storage at 4 °C and 37 °C was shown to induce a size decline and degradation of exosomes, while repeated freeze cycles at −20°C and thawing did not alter the behavior of exosomes [118].

In an ISEV (International Society of Extracellular Vesicles) position paper that summarizes discussions that took place at the ISEV research workshop in New York in 2012, the recommendations for the storage of EVs include the use of siliconized vessels to prevent the adherence of EVs to batch surfaces. Furthermore, the suggested resuspension medium is PBS and the best storage temperature is considered to be −80 °C. Interestingly, neither the freeze and thawing cycles, nor the osmolarity environment, seem to affect the stability of EVs [119].

An investigation on EVs derived from neutrophilic granulocytes confirmed a better storage stability at −80 °C for 4 weeks. In contrast, EVs reduced in size and number at −20 °C and 4 °C, respectively [120].

The basic nature of biobanking requires the study of protocol standardization, including precise evaluation of samples in terms of the harvesting rate, characterization, pre-clinical parameter assessment, factors that could affect the long-term storage of samples, database archiving, and distribution. In addition, complete information regarding the life cycle of the exosomes is critical for the understanding of sample integrity and quality.

More investigations are needed to bridge this gap of knowledge and to create standardized GMP (Good-Manufacturing Practice) protocols for safe and effective exosome cryopreservation and storage, in order to obtain optimal quality samples for applications in the most disparate clinical situations.

6. Conclusions

The almost infinite power of the MSC secretome is currently well known, as demonstrated by the growing number of clinical applications involving these cells. The discovery that extracellular vesicles budding from MSCs conserve and spread precious biological information, allows many of the regulatory issues linked to cell therapy to be overcome.

The previously proposed novel "NANOBIOME" approach is based on the biobanking of EVs secreted by MSCs for their easy and available storage and distribution. The standardization of protocols for the isolation and cryopreservation of MSC exosomes could represent a very attractive and useful topic for regenerative medicine purposes.

Author Contributions: Conceptualization, M.T. and B.C.; Methodology, F.P. and M.M.; Validation, M.T., B.C., M.M. and F.P.; Data Curation, B.C.; Writing-Original Draft Preparation, B.C. and M.T.; Writing-Review & Editing, M.T., B.C., M.M. and F.P.; Supervision, M.T.; Funding Acquisition, M.T. and M.M.

Funding: This research was funded by the following project—"ZEUSI—customiZEd prodUcts applied to innovative proceSses in regenerative and reconstructIve surgery of maxillofacial region (nuovi prodotti custom-made applicati a processi innovativi utilizzati nella chirurgia rigenerativa e ricostruttiva della regione maxillo-facciale)"; grant number: MISE F/050160/00/X32.

Conflicts of Interest: The authors declare no conflict of interest.

References

1. Tatullo, M.; Marrelli, M.; Paduano, F. The regenerative medicine in oral and maxillofacial surgery: The most important innovations in the clinical application of mesenchymal stem cells. *Int. J. Med. Sci.* **2015**, *12*, 72–77. [CrossRef] [PubMed]
2. Tatullo, M.; Codispoti, B.; Pacifici, A.; Palmieri, F.; Marrelli, M.; Pacifici, L.; Paduano, F. Potential use of human periapical cyst-mesenchymal stem cells (hPCy-MSCs) as a novel stem cell source for regenerative medicine applications. *Front. Cell Dev. Biol.* **2017**, *5*, 103. [CrossRef] [PubMed]
3. Tatullo, M.; Falisi, G.; Amantea, M.; Rastelli, C.; Paduano, F.; Marrelli, M. Dental pulp stem cells and human periapical cyst mesenchymal stem cells in bone tissue regeneration: Comparison of basal and osteogenic differentiated gene expression of a newly discovered mesenchymal stem cell lineage. *J. Biol. Regul. Homeost. Agents* **2015**, *29*, 713–718. [PubMed]
4. Marrelli, M.; Paduano, F.; Tatullo, M. Human periapical cyst-mesenchymal stem cells differentiate into neuronal cells. *J. Dent. Res.* **2015**, *94*, 843–852. [CrossRef] [PubMed]
5. Ryan, J.M.; Barry, F.P.; Murphy, J.M.; Mahon, B.P. Mesenchymal stem cells avoid allogeneic rejection. *J. Inflamm.* **2005**, *2*, 8. [CrossRef] [PubMed]

6. Le Blanc, K.; Tammik, L.; Sundberg, B.; Haynesworth, S.E.; Ringden, O. Mesenchymal stem cells inhibit and stimulate mixed lymphocyte cultures and mitogenic responses independently of the major histocompatibility complex. *Scand. J. Immunol.* **2003**, *57*, 11–20. [CrossRef] [PubMed]
7. Liang, X.; Ding, Y.; Zhang, Y.; Tse, H.F.; Lian, Q. Paracrine mechanisms of mesenchymal stem cell-based therapy: Current status and perspectives. *Cell Transplant.* **2014**, *23*, 1045–1059. [CrossRef] [PubMed]
8. Jeyaram, A.; Jay, S.M. Preservation and storage stability of extracellular vesicles for therapeutic applications. *AAPS J.* **2017**, *20*, 1. [CrossRef] [PubMed]
9. de Miguel-Beriain, I. The ethics of stem cells revisited. *Adv. Drug Deliv. Rev.* **2015**, *82–83*, 176–180. [CrossRef] [PubMed]
10. Chen, T.; Wang, X.; Guo, L.; Wu, M.; Duan, Z.; Lv, J.; Tai, W.; Renganathan, H.; Didier, R.; Li, J.; et al. Embryonic stem cells promoting macrophage survival and function are crucial for teratoma development. *Front. Immunol.* **2014**, *5*, 275. [CrossRef] [PubMed]
11. Dricu, A. Recent challenges with stem cell banking. *Expert Opin. Biol. Ther.* **2018**, *18*, 355–358. [CrossRef] [PubMed]
12. Inamoto, Y.; Shah, N.N.; Savani, B.N.; Shaw, B.E.; Abraham, A.A.; Ahmed, I.A.; Akpek, G.; Atsuta, Y.; Baker, K.S.; Basak, G.W.; et al. Secondary solid cancer screening following hematopoietic cell transplantation. *Bone Marrow Transplant.* **2015**, *50*, 1013–1023. [CrossRef] [PubMed]
13. Ehrhardt, M.J.; Brazauskas, R.; He, W.; Rizzo, J.D.; Shaw, B.E. Survival of patients who develop solid tumors following hematopoietic stem cell transplantation. *Bone Marrow Transplant.* **2016**, *51*, 83–88. [CrossRef] [PubMed]
14. Harris, D.T. Banking of adipose- and cord tissue-derived stem cells: Technical and regulatory issues. *Adv. Exp. Med. Biol.* **2016**, *951*, 147–154. [PubMed]
15. Codispoti, B.; Rinaldo, N.; Chiarella, E.; Lupia, M.; Spoleti, C.B.; Marafioti, M.G.; Aloisio, A.; Scicchitano, S.; Giordano, M.; Nappo, G.; et al. Recombinant TAT-BMI-1 fusion protein induces ex vivo expansion of human umbilical cord blood-derived hematopoietic stem cells. *Oncotarget* **2017**, *8*, 43782–43798. [CrossRef] [PubMed]
16. Taylor-Weiner, H.; Graff Zivin, J. Medicine's wild west—Unlicensed stem-cell clinics in the united states. *N. Engl. J. Med.* **2015**, *373*, 985–987. [CrossRef] [PubMed]
17. Thirumala, S.; Goebel, W.S.; Woods, E.J. Manufacturing and banking of mesenchymal stem cells. *Expert Opin. Biol. Ther.* **2013**, *13*, 673–691. [CrossRef] [PubMed]
18. Raposo, G.; Stoorvogel, W. Extracellular vesicles: Exosomes, microvesicles, and friends. *J. Cell Biol.* **2013**, *200*, 373–383. [CrossRef] [PubMed]
19. Colombo, M.; Raposo, G.; Thery, C. Biogenesis, secretion, and intercellular interactions of exosomes and other extracellular vesicles. *Annu. Rev. Cell Dev. Biol.* **2014**, *30*, 255–289. [CrossRef] [PubMed]
20. Gould, S.J.; Raposo, G. As we wait: Coping with an imperfect nomenclature for extracellular vesicles. *J. Extracell. Vesicles* **2013**, *2*, 20389. [CrossRef] [PubMed]
21. Cocucci, E.; Meldolesi, J. Ectosomes and exosomes: Shedding the confusion between extracellular vesicles. *Trends Cell Biol.* **2015**, *25*, 364–372. [CrossRef] [PubMed]
22. Pan, B.T.; Johnstone, R.M. Fate of the transferrin receptor during maturation of sheep reticulocytes in vitro: Selective externalization of the receptor. *Cell* **1983**, *33*, 967–978. [CrossRef]
23. Johnstone, R.M.; Adam, M.; Hammond, J.R.; Orr, L.; Turbide, C. Vesicle formation during reticulocyte maturation. Association of plasma membrane activities with released vesicles (exosomes). *J. Biol. Chem.* **1987**, *262*, 9412–9420. [PubMed]
24. Trams, E.G.; Lauter, C.J.; Salem, N., Jr.; Heine, U. Exfoliation of membrane ecto-enzymes in the form of micro-vesicles. *Biochim. Biophys. Acta* **1981**, *645*, 63–70. [CrossRef]
25. Harding, C.; Heuser, J.; Stahl, P. Receptor-mediated endocytosis of transferrin and recycling of the transferrin receptor in rat reticulocytes. *J. Cell Biol.* **1983**, *97*, 329–339. [CrossRef] [PubMed]
26. Raposo, G.; Nijman, H.W.; Stoorvogel, W.; Liejendekker, R.; Harding, C.V.; Melief, C.J.; Geuze, H.J. B lymphocytes secrete antigen-presenting vesicles. *J. Exp. Med.* **1996**, *183*, 1161–1172. [CrossRef] [PubMed]
27. Satta, N.; Toti, F.; Feugeas, O.; Bohbot, A.; Dachary-Prigent, J.; Eschwege, V.; Hedman, H.; Freyssinet, J.M. Monocyte vesiculation is a possible mechanism for dissemination of membrane-associated procoagulant activities and adhesion molecules after stimulation by lipopolysaccharide. *J. Immunol.* **1994**, *153*, 3245–3255. [PubMed]

28. Sims, P.J.; Wiedmer, T.; Esmon, C.T.; Weiss, H.J.; Shattil, S.J. Assembly of the platelet prothrombinase complex is linked to vesiculation of the platelet plasma membrane. Studies in Scott syndrome: An isolated defect in platelet procoagulant activity. *J. Biol. Chem.* **1989**, *264*, 17049–17057. [PubMed]
29. Mesri, M.; Altieri, D.C. Endothelial cell activation by leukocyte microparticles. *J. Immunol.* **1998**, *161*, 4382–4387. [PubMed]
30. Baj-Krzyworzeka, M.; Majka, M.; Pratico, D.; Ratajczak, J.; Vilaire, G.; Kijowski, J.; Reca, R.; Janowska-Wieczorek, A.; Ratajczak, M.Z. Platelet-derived microparticles stimulate proliferation, survival, adhesion, and chemotaxis of hematopoietic cells. *Exp. Hematol.* **2002**, *30*, 450–459. [CrossRef]
31. Poutsiaka, D.D.; Schroder, E.W.; Taylor, D.D.; Levy, E.M.; Black, P.H. Membrane vesicles shed by murine melanoma cells selectively inhibit the expression of Ia antigen by macrophages. *J. Immunol.* **1985**, *134*, 138–144. [PubMed]
32. Zhang, H.G.; Grizzle, W.E. Exosomes and cancer: A newly described pathway of immune suppression. *Clin. Cancer Res.* **2011**, *17*, 959–964. [CrossRef] [PubMed]
33. Wolfers, J.; Lozier, A.; Raposo, G.; Regnault, A.; Thery, C.; Masurier, C.; Flament, C.; Pouzieux, S.; Faure, F.; Tursz, T.; et al. Tumor-derived exosomes are a source of shared tumor rejection antigens for CTL cross-priming. *Nat. Med.* **2001**, *7*, 297–303. [CrossRef] [PubMed]
34. Sheldon, H.; Heikamp, E.; Turley, H.; Dragovic, R.; Thomas, P.; Oon, C.E.; Leek, R.; Edelmann, M.; Kessler, B.; Sainson, R.C.; et al. New mechanism for notch signaling to endothelium at a distance by delta-like 4 incorporation into exosomes. *Blood* **2010**, *116*, 2385–2394. [CrossRef] [PubMed]
35. Peinado, H.; Aleckovic, M.; Lavotshkin, S.; Matei, I.; Costa-Silva, B.; Moreno-Bueno, G.; Hergueta-Redondo, M.; Williams, C.; Garcia-Santos, G.; Ghajar, C.; et al. Melanoma exosomes educate bone marrow progenitor cells toward a pro-metastatic phenotype through met. *Nat. Med.* **2012**, *18*, 883–891. [CrossRef] [PubMed]
36. Kalra, H.; Simpson, R.J.; Ji, H.; Aikawa, E.; Altevogt, P.; Askenase, P.; Bond, V.C.; Borras, F.E.; Breakefield, X.; Budnik, V.; et al. Vesiclepedia: A compendium for extracellular vesicles with continuous community annotation. *PLoS Biol.* **2012**, *10*, e1001450. [CrossRef] [PubMed]
37. Kim, D.K.; Kang, B.; Kim, O.Y.; Choi, D.S.; Lee, J.; Kim, S.R.; Go, G.; Yoon, Y.J.; Kim, J.H.; Jang, S.C.; et al. Evpedia: An integrated database of high-throughput data for systemic analyses of extracellular vesicles. *J. Extracell. Vesicles* **2013**, *2*, 20384. [CrossRef] [PubMed]
38. Thery, C.; Amigorena, S.; Raposo, G.; Clayton, A. Isolation and characterization of exosomes from cell culture supernatants and biological fluids. *Curr. Protoc. Cell Biol.* **2006**, *3*, 3–22. [CrossRef] [PubMed]
39. Baietti, M.F.; Zhang, Z.; Mortier, E.; Melchior, A.; Degeest, G.; Geeraerts, A.; Ivarsson, Y.; Depoortere, F.; Coomans, C.; Vermeiren, E.; et al. Syndecan-syntenin-alix regulates the biogenesis of exosomes. *Nat. Cell Biol.* **2012**, *14*, 677–685. [CrossRef] [PubMed]
40. Taylor, D.D.; Homesley, H.D.; Doellgast, G.J. "Membrane-associated" immunoglobulins in cyst and ascites fluids of ovarian cancer patients. *Am. J. Reprod. Immunol.* **1983**, *3*, 7–11. [CrossRef] [PubMed]
41. Lamparski, H.G.; Metha-Damani, A.; Yao, J.Y.; Patel, S.; Hsu, D.H.; Ruegg, C.; Le Pecq, J.B. Production and characterization of clinical grade exosomes derived from dendritic cells. *J. Immunol. Methods* **2002**, *270*, 211–226. [CrossRef]
42. Dragovic, R.A.; Gardiner, C.; Brooks, A.S.; Tannetta, D.S.; Ferguson, D.J.; Hole, P.; Carr, B.; Redman, C.W.; Harris, A.L.; Dobson, P.J.; et al. Sizing and phenotyping of cellular vesicles using nanoparticle tracking analysis. *Nanomedicine* **2011**, *7*, 780–788. [CrossRef] [PubMed]
43. Di Vizio, D.; Morello, M.; Dudley, A.C.; Schow, P.W.; Adam, R.M.; Morley, S.; Mulholland, D.; Rotinen, M.; Hager, M.H.; Insabato, L.; et al. Large oncosomes in human prostate cancer tissues and in the circulation of mice with metastatic disease. *Am. J. Pathol.* **2012**, *181*, 1573–1584. [CrossRef] [PubMed]
44. Lacroix, R.; Judicone, C.; Mooberry, M.; Boucekine, M.; Key, N.S.; Dignat-George, F.; The, I.S.S.C.W. Standardization of pre-analytical variables in plasma microparticle determination: Results of the International Society on Thrombosis and Haemostasis SSC Collaborative Workshop. *J. Thromb. Haemost.* **2013**, *11*, 1190–1193. [CrossRef] [PubMed]
45. Llorente, A.; Skotland, T.; Sylvanne, T.; Kauhanen, D.; Rog, T.; Orlowski, A.; Vattulainen, I.; Ekroos, K.; Sandvig, K. Molecular lipidomics of exosomes released by pc-3 prostate cancer cells. *Biochim. Biophys. Acta* **2013**, *1831*, 1302–1309. [CrossRef] [PubMed]

46. Dachary-Prigent, J.; Freyssinet, J.M.; Pasquet, J.M.; Carron, J.C.; Nurden, A.T. Annexin V as a probe of aminophospholipid exposure and platelet membrane vesiculation: A flow cytometry study showing a role for free sulfhydryl groups. *Blood* **1993**, *81*, 2554–2565. [PubMed]
47. Thery, C.; Boussac, M.; Veron, P.; Ricciardi-Castagnoli, P.; Raposo, G.; Garin, J.; Amigorena, S. Proteomic analysis of dendritic cell-derived exosomes: A secreted subcellular compartment distinct from apoptotic vesicles. *J. Immunol.* **2001**, *166*, 7309–7318. [CrossRef] [PubMed]
48. Valadi, H.; Ekstrom, K.; Bossios, A.; Sjostrand, M.; Lee, J.J.; Lotvall, J.O. Exosome-mediated transfer of mrnas and micrornas is a novel mechanism of genetic exchange between cells. *Nat. Cell Biol.* **2007**, *9*, 654–659. [CrossRef] [PubMed]
49. Hunter, M.P.; Ismail, N.; Zhang, X.; Aguda, B.D.; Lee, E.J.; Yu, L.; Xiao, T.; Schafer, J.; Lee, M.L.; Schmittgen, T.D.; et al. Detection of microrna expression in human peripheral blood microvesicles. *PLoS ONE* **2008**, *3*, e3694. [CrossRef] [PubMed]
50. Skog, J.; Wurdinger, T.; van Rijn, S.; Meijer, D.H.; Gainche, L.; Sena-Esteves, M.; Curry, W.T., Jr.; Carter, B.S.; Krichevsky, A.M.; Breakefield, X.O. Glioblastoma microvesicles transport rna and proteins that promote tumour growth and provide diagnostic biomarkers. *Nat. Cell Biol.* **2008**, *10*, 1470–1476. [CrossRef] [PubMed]
51. Nolte-'t Hoen, E.N.; Buermans, H.P.; Waasdorp, M.; Stoorvogel, W.; Wauben, M.H.; t Hoen, P.A. Deep sequencing of rna from immune cell-derived vesicles uncovers the selective incorporation of small non-coding RNA biotypes with potential regulatory functions. *Nucleic Acids Res.* **2012**, *40*, 9272–9285. [CrossRef] [PubMed]
52. Kilpinen, L.; Impola, U.; Sankkila, L.; Ritamo, I.; Aatonen, M.; Kilpinen, S.; Tuimala, J.; Valmu, L.; Levijoki, J.; Finckenberg, P.; et al. Extracellular membrane vesicles from umbilical cord blood-derived MSC protect against ischemic acute kidney injury, a feature that is lost after inflammatory conditioning. *J. Extracell. Vesicles* **2013**, *2*, 21927. [CrossRef] [PubMed]
53. de Jong, O.G.; Verhaar, M.C.; Chen, Y.; Vader, P.; Gremmels, H.; Posthuma, G.; Schiffelers, R.M.; Gucek, M.; van Balkom, B.W. Cellular stress conditions are reflected in the protein and RNA content of endothelial cell-derived exosomes. *J. Extracell. Vesicles* **2012**, *1*, 18396. [CrossRef] [PubMed]
54. Parolini, I.; Federici, C.; Raggi, C.; Lugini, L.; Palleschi, S.; De Milito, A.; Coscia, C.; Iessi, E.; Logozzi, M.; Molinari, A.; et al. Microenvironmental pH is a key factor for exosome traffic in tumor cells. *J. Biol. Chem.* **2009**, *284*, 34211–34222. [CrossRef] [PubMed]
55. Toh, W.S.; Lai, R.C.; Zhang, B.; Lim, S.K. MSC exosome works through a protein-based mechanism of action. *Biochem. Soc. Trans.* **2018**, *46*, 843–853. [CrossRef] [PubMed]
56. Gould, G.W.; Lippincott-Schwartz, J. New roles for endosomes: From vesicular carriers to multi-purpose platforms. *Nat. Rev. Mol. Cell Biol.* **2009**, *10*, 287–292. [CrossRef] [PubMed]
57. Stuffers, S.; Sem Wegner, C.; Stenmark, H.; Brech, A. Multivesicular endosome biogenesis in the absence of ESCRTs. *Traffic* **2009**, *10*, 925–937. [CrossRef] [PubMed]
58. Hanson, P.I.; Cashikar, A. Multivesicular body morphogenesis. *Annu. Rev. Cell Dev. Biol.* **2012**, *28*, 337–362. [CrossRef] [PubMed]
59. van Niel, G.; Charrin, S.; Simoes, S.; Romao, M.; Rochin, L.; Saftig, P.; Marks, M.S.; Rubinstein, E.; Raposo, G. The tetraspanin CD63 regulates ESCRT-independent and -dependent endosomal sorting during melanogenesis. *Dev. Cell* **2011**, *21*, 708–721. [CrossRef] [PubMed]
60. Trajkovic, K.; Hsu, C.; Chiantia, S.; Rajendran, L.; Wenzel, D.; Wieland, F.; Schwille, P.; Brugger, B.; Simons, M. Ceramide triggers budding of exosome vesicles into multivesicular endosomes. *Science* **2008**, *319*, 1244–1247. [CrossRef] [PubMed]
61. Ghossoub, R.; Lembo, F.; Rubio, A.; Gaillard, C.B.; Bouchet, J.; Vitale, N.; Slavik, J.; Machala, M.; Zimmermann, P. Syntenin-alix exosome biogenesis and budding into multivesicular bodies are controlled by ARF6 and PLD2. *Nat. Commun.* **2014**, *5*, 3477. [CrossRef] [PubMed]
62. Pasquet, J.M.; Dachary-Prigent, J.; Nurden, A.T. Calcium influx is a determining factor of calpain activation and microparticle formation in platelets. *Eur. J. Biochem.* **1996**, *239*, 647–654. [CrossRef] [PubMed]
63. Buschow, S.I.; Nolte-'t Hoen, E.N.; van Niel, G.; Pols, M.S.; ten Broeke, T.; Lauwen, M.; Ossendorp, F.; Melief, C.J.; Raposo, G.; Wubbolts, R.; et al. MHC II in dendritic cells is targeted to lysosomes or T cell-induced exosomes via distinct multivesicular body pathways. *Traffic* **2009**, *10*, 1528–1542. [CrossRef] [PubMed]

64. Fruhbeis, C.; Frohlich, D.; Kuo, W.P.; Amphornrat, J.; Thilemann, S.; Saab, A.S.; Kirchhoff, F.; Mobius, W.; Goebbels, S.; Nave, K.A.; et al. Neurotransmitter-triggered transfer of exosomes mediates oligodendrocyte-neuron communication. *PLoS Biol.* **2013**, *11*, e1001604. [CrossRef] [PubMed]
65. Stenmark, H. Rab GTPases as coordinators of vesicle traffic. *Nat. Rev. Mol. Cell Biol.* **2009**, *10*, 513–525. [CrossRef] [PubMed]
66. Zylbersztejn, K.; Galli, T. Vesicular traffic in cell navigation. *FEBS J.* **2011**, *278*, 4497–4505. [CrossRef] [PubMed]
67. Morelli, A.E.; Larregina, A.T.; Shufesky, W.J.; Sullivan, M.L.; Stolz, D.B.; Papworth, G.D.; Zahorchak, A.F.; Logar, A.J.; Wang, Z.; Watkins, S.C.; et al. Endocytosis, intracellular sorting, and processing of exosomes by dendritic cells. *Blood* **2004**, *104*, 3257–3266. [CrossRef] [PubMed]
68. Christianson, H.C.; Svensson, K.J.; van Kuppevelt, T.H.; Li, J.P.; Belting, M. Cancer cell exosomes depend on cell-surface heparan sulfate proteoglycans for their internalization and functional activity. *Proc. Natl. Acad. Sci. USA* **2013**, *110*, 17380–17385. [CrossRef] [PubMed]
69. Abrami, L.; Brandi, L.; Moayeri, M.; Brown, M.J.; Krantz, B.A.; Leppla, S.H.; van der Goot, F.G. Hijacking multivesicular bodies enables long-term and exosome-mediated long-distance action of anthrax toxin. *Cell Rep.* **2013**, *5*, 986–996. [CrossRef] [PubMed]
70. Yu, B.; Zhang, X.; Li, X. Exosomes derived from mesenchymal stem cells. *Int. J. Mol. Sci.* **2014**, *15*, 4142–4157. [CrossRef] [PubMed]
71. Bruno, S.; Grange, C.; Deregibus, M.C.; Calogero, R.A.; Saviozzi, S.; Collino, F.; Morando, L.; Busca, A.; Falda, M.; Bussolati, B.; et al. Mesenchymal stem cell-derived microvesicles protect against acute tubular injury. *J. Am. Soc. Nephrol.* **2009**, *20*, 1053–1067. [CrossRef] [PubMed]
72. Lai, R.C.; Arslan, F.; Lee, M.M.; Sze, N.S.; Choo, A.; Chen, T.S.; Salto-Tellez, M.; Timmers, L.; Lee, C.N.; El Oakley, R.M.; et al. Exosome secreted by MSC reduces myocardial ischemia/reperfusion injury. *Stem Cell Res.* **2010**, *4*, 214–222. [CrossRef] [PubMed]
73. Ramos, T.L.; Sanchez-Abarca, L.I.; Muntion, S.; Preciado, S.; Puig, N.; Lopez-Ruano, G.; Hernandez-Hernandez, A.; Redondo, A.; Ortega, R.; Rodriguez, C.; et al. MSC surface markers (CD44, CD 73, and CD 90) can identify human MSC-derived extracellular vesicles by conventional flow cytometry. *Cell Commun. Signal.* **2016**, *14*, 2. [CrossRef] [PubMed]
74. Lai, R.C.; Tan, S.S.; Teh, B.J.; Sze, S.K.; Arslan, F.; de Kleijn, D.P.; Choo, A.; Lim, S.K. Proteolytic potential of the MSC exosome proteome: Implications for an exosome-mediated delivery of therapeutic proteasome. *Int J. Proteom.* **2012**, *2012*, 971907. [CrossRef] [PubMed]
75. Chen, T.S.; Lai, R.C.; Lee, M.M.; Choo, A.B.; Lee, C.N.; Lim, S.K. Mesenchymal stem cell secretes microparticles enriched in pre-micrornas. *Nucleic Acids Res.* **2010**, *38*, 215–224. [CrossRef] [PubMed]
76. Bian, S.; Zhang, L.; Duan, L.; Wang, X.; Min, Y.; Yu, H. Extracellular vesicles derived from human bone marrow mesenchymal stem cells promote angiogenesis in a rat myocardial infarction model. *J. Mol. Med.* **2014**, *92*, 387–397. [CrossRef] [PubMed]
77. Zhou, Y.; Xu, H.; Xu, W.; Wang, B.; Wu, H.; Tao, Y.; Zhang, B.; Wang, M.; Mao, F.; Yan, Y.; et al. Exosomes released by human umbilical cord mesenchymal stem cells protect against cisplatin-induced renal oxidative stress and apoptosis in vivo and in vitro. *Stem Cell Res. Ther.* **2013**, *4*, 34. [CrossRef] [PubMed]
78. Ranjit, S.; Patters, B.J.; Gerth, K.A.; Haque, S.; Choudhary, S.; Kumar, S. Potential neuroprotective role of astroglial exosomes against smoking-induced oxidative stress and HIV-1 replication in the central nervous system. *Expert Opin. Ther. Targets* **2018**, *22*, 703–714. [CrossRef] [PubMed]
79. Bucan, V.; Vaslaitis, D.; Peck, C.T.; Strauss, S.; Vogt, P.M.; Radtke, C. Effect of exosomes from rat adipose-derived mesenchymal stem cells on neurite outgrowth and sciatic nerve regeneration after crush injury. *Mol. Neurobiol.* **2018**. [CrossRef] [PubMed]
80. Cai, Z.Y.; Xiao, M.; Quazi, S.H.; Ke, Z.Y. Exosomes: A novel therapeutic target for Alzheimer's disease? *Neural Regen. Res.* **2018**, *13*, 930–935. [CrossRef] [PubMed]
81. Hervera, A.; De Virgiliis, F.; Palmisano, I.; Zhou, L.; Tantardini, E.; Kong, G.; Hutson, T.; Danzi, M.C.; Perry, R.B.; Santos, C.X.C.; et al. Reactive oxygen species regulate axonal regeneration through the release of exosomal NADPH oxidase 2 complexes into injured axons. *Nat. Cell Biol.* **2018**, *20*, 307–319. [CrossRef] [PubMed]

82. Nakamura, Y.; Miyaki, S.; Ishitobi, H.; Matsuyama, S.; Nakasa, T.; Kamei, N.; Akimoto, T.; Higashi, Y.; Ochi, M. Mesenchymal-stem-cell-derived exosomes accelerate skeletal muscle regeneration. *FEBS Lett.* **2015**, *589*, 1257–1265. [CrossRef] [PubMed]
83. Damania, A.; Jaiman, D.; Teotia, A.K.; Kumar, A. Mesenchymal stromal cell-derived exosome-rich fractionated secretome confers a hepatoprotective effect in liver injury. *Stem Cell Res. Ther.* **2018**, *9*, 31. [CrossRef] [PubMed]
84. Ranghino, A.; Bruno, S.; Bussolati, B.; Moggio, A.; Dimuccio, V.; Tapparo, M.; Biancone, L.; Gontero, P.; Frea, B.; Camussi, G. The effects of glomerular and tubular renal progenitors and derived extracellular vesicles on recovery from acute kidney injury. *Stem Cell Res. Ther.* **2017**, *8*, 24. [CrossRef] [PubMed]
85. Collino, F.; Bruno, S.; Incarnato, D.; Dettori, D.; Neri, F.; Provero, P.; Pomatto, M.; Oliviero, S.; Tetta, C.; Quesenberry, P.J.; et al. AKI recovery induced by mesenchymal stromal cell-derived extracellular vesicles carrying micrornas. *J. Am. Soc. Nephrol.* **2015**, *26*, 2349–2360. [CrossRef] [PubMed]
86. Ma, J.; Zhao, Y.; Sun, L.; Sun, X.; Zhao, X.; Sun, X.; Qian, H.; Xu, W.; Zhu, W. Exosomes derived from AKT-modified human umbilical cord mesenchymal stem cells improve cardiac regeneration and promote angiogenesis via activating platelet-derived growth factor D. *Stem Cells Transl. Med.* **2017**, *6*, 51–59. [CrossRef] [PubMed]
87. Yu, B.; Kim, H.W.; Gong, M.; Wang, J.; Millard, R.W.; Wang, Y.; Ashraf, M.; Xu, M. Exosomes secreted from GATA-4 overexpressing mesenchymal stem cells serve as a reservoir of anti-apoptotic micrornas for cardioprotection. *Int. J. Cardiol.* **2015**, *182*, 349–360. [CrossRef] [PubMed]
88. Sze, S.K.; de Kleijn, D.P.; Lai, R.C.; Khia Way Tan, E.; Zhao, H.; Yeo, K.S.; Low, T.Y.; Lian, Q.; Lee, C.N.; Mitchell, W.; et al. Elucidating the secretion proteome of human embryonic stem cell-derived mesenchymal stem cells. *Mol. Cell. Proteom.* **2007**, *6*, 1680–1689. [CrossRef] [PubMed]
89. Wang, Y.; Yu, D.; Liu, Z.; Zhou, F.; Dai, J.; Wu, B.; Zhou, J.; Heng, B.C.; Zou, X.H.; Ouyang, H.; et al. Exosomes from embryonic mesenchymal stem cells alleviate osteoarthritis through balancing synthesis and degradation of cartilage extracellular matrix. *Stem Cell Res. Ther.* **2017**, *8*, 189. [CrossRef] [PubMed]
90. Zhang, S.; Chu, W.C.; Lai, R.C.; Lim, S.K.; Hui, J.H.; Toh, W.S. Exosomes derived from human embryonic mesenchymal stem cells promote osteochondral regeneration. *Osteoarthr. Cartil.* **2016**, *24*, 2135–2140. [CrossRef] [PubMed]
91. Lai, R.C.; Yeo, R.W.; Padmanabhan, J.; Choo, A.; de Kleijn, D.P.; Lim, S.K. Isolation and characterization of exosome from human embryonic stem cell-derived C-Myc-immortalized mesenchymal stem cells. *Methods Mol. Biol.* **2016**, *1416*, 477–494. [PubMed]
92. Hu, G.W.; Li, Q.; Niu, X.; Hu, B.; Liu, J.; Zhou, S.M.; Guo, S.C.; Lang, H.L.; Zhang, C.Q.; Wang, Y.; et al. Exosomes secreted by human-induced pluripotent stem cell-derived mesenchymal stem cells attenuate limb ischemia by promoting angiogenesis in mice. *Stem Cell Res. Ther.* **2015**, *6*, 10. [CrossRef] [PubMed]
93. Zhang, J.; Guan, J.; Niu, X.; Hu, G.; Guo, S.; Li, Q.; Xie, Z.; Zhang, C.; Wang, Y. Exosomes released from human induced pluripotent stem cells-derived MSCs facilitate cutaneous wound healing by promoting collagen synthesis and angiogenesis. *J. Transl. Med.* **2015**, *13*, 49. [CrossRef] [PubMed]
94. Liu, X.; Li, Q.; Niu, X.; Hu, B.; Chen, S.; Song, W.; Ding, J.; Zhang, C.; Wang, Y. Exosomes secreted from human-induced pluripotent stem cell-derived mesenchymal stem cells prevent osteonecrosis of the femoral head by promoting angiogenesis. *Int. J. Biol. Sci.* **2017**, *13*, 232–244. [CrossRef] [PubMed]
95. Budoni, M.; Fierabracci, A.; Luciano, R.; Petrini, S.; Di Ciommo, V.; Muraca, M. The immunosuppressive effect of mesenchymal stromal cells on B lymphocytes is mediated by membrane vesicles. *Cell Transplant.* **2013**, *22*, 369–379. [CrossRef] [PubMed]
96. Mokarizadeh, A.; Delirezh, N.; Morshedi, A.; Mosayebi, G.; Farshid, A.A.; Mardani, K. Microvesicles derived from mesenchymal stem cells: Potent organelles for induction of tolerogenic signaling. *Immunol. Lett.* **2012**, *147*, 47–54. [CrossRef] [PubMed]
97. Zhang, B.; Yin, Y.; Lai, R.C.; Tan, S.S.; Choo, A.B.; Lim, S.K. Mesenchymal stem cells secrete immunologically active exosomes. *Stem Cells Dev.* **2014**, *23*, 1233–1244. [CrossRef] [PubMed]
98. Lo Sicco, C.; Reverberi, D.; Balbi, C.; Ulivi, V.; Principi, E.; Pascucci, L.; Becherini, P.; Bosco, M.C.; Varesio, L.; Franzin, C.; et al. Mesenchymal stem cell-derived extracellular vesicles as mediators of anti-inflammatory effects: Endorsement of macrophage polarization. *Stem Cells Transl. Med.* **2017**, *6*, 1018–1028. [CrossRef] [PubMed]

99. Eirin, A.; Zhu, X.Y.; Puranik, A.S.; Tang, H.; McGurren, K.A.; van Wijnen, A.J.; Lerman, A.; Lerman, L.O. Mesenchymal stem cell-derived extracellular vesicles attenuate kidney inflammation. *Kidney Int.* **2017**, *92*, 114–124. [CrossRef] [PubMed]
100. Du, Y.M.; Zhuansun, Y.X.; Chen, R.; Lin, L.; Lin, Y.; Li, J.G. Mesenchymal stem cell exosomes promote immunosuppression of regulatory T cells in asthma. *Exp. Cell Res.* **2018**, *363*, 114–120. [CrossRef] [PubMed]
101. Li, T.; Yan, Y.; Wang, B.; Qian, H.; Zhang, X.; Shen, L.; Wang, M.; Zhou, Y.; Zhu, W.; Li, W.; et al. Exosomes derived from human umbilical cord mesenchymal stem cells alleviate liver fibrosis. *Stem Cells Dev.* **2013**, *22*, 845–854. [CrossRef] [PubMed]
102. Shen, B.; Liu, J.; Zhang, F.; Wang, Y.; Qin, Y.; Zhou, Z.; Qiu, J.; Fan, Y. CCR2 positive exosome released by mesenchymal stem cells suppresses macrophage functions and alleviates ischemia/reperfusion-induced renal injury. *Stem Cells Int.* **2016**, *2016*, 1240301. [CrossRef] [PubMed]
103. Liu, W.; Wang, Y.; Gong, F.; Rong, Y.; Luo, Y.; Tang, P.; Zhou, Z.; Zhou, Z.; Xu, T.; Jiang, T.; et al. Exosomes derived from bone mesenchymal stem cells repair traumatic spinal cord injury by suppressing the activation of A1 neurotoxic reactive astrocytes. *J. Neurotrauma* **2018**. [CrossRef] [PubMed]
104. Sun, G.; Li, G.; Li, D.; Huang, W.; Zhang, R.; Zhang, H.; Duan, Y.; Wang, B. HUCMSC derived exosomes promote functional recovery in spinal cord injury mice via attenuating inflammation. *Mater. Sci. Eng. C Mater. Biol. Appl.* **2018**, *89*, 194–204. [CrossRef] [PubMed]
105. Caruso, S.; Poon, I.K.H. Apoptotic cell-derived extracellular vesicles: More than just debris. *Front. Immunol.* **2018**, *9*, 1486. [CrossRef] [PubMed]
106. Zirngibl, M.; Furnrohr, B.G.; Janko, C.; Munoz, L.E.; Voll, R.E.; Gregory, C.D.; Schett, G.; Herrmann, M. Loading of nuclear autoantigens prototypically recognized by systemic lupus erythematosus sera into late apoptotic vesicles requires intact microtubules and myosin light chain kinase activity. *Clin. Exp. Immunol.* **2015**, *179*, 39–49. [CrossRef] [PubMed]
107. Nagata, S.; Hanayama, R.; Kawane, K. Autoimmunity and the clearance of dead cells. *Cell* **2010**, *140*, 619–630. [CrossRef] [PubMed]
108. Muhsin-Sharafaldine, M.R.; Saunderson, S.C.; Dunn, A.C.; Faed, J.M.; Kleffmann, T.; McLellan, A.D. Procoagulant and immunogenic properties of melanoma exosomes, microvesicles and apoptotic vesicles. *Oncotarget* **2016**, *7*, 56279–56294. [CrossRef] [PubMed]
109. Winau, F.; Weber, S.; Sad, S.; de Diego, J.; Hoops, S.L.; Breiden, B.; Sandhoff, K.; Brinkmann, V.; Kaufmann, S.H.; Schaible, U.E. Apoptotic vesicles crossprime CD8 T cells and protect against tuberculosis. *Immunity* **2006**, *24*, 105–117. [CrossRef] [PubMed]
110. Properzi, F.; Logozzi, M.; Fais, S. Exosomes: The future of biomarkers in medicine. *Biomark. Med.* **2013**, *7*, 769–778. [CrossRef] [PubMed]
111. Vader, P.; Breakefield, X.O.; Wood, M.J. Extracellular vesicles: Emerging targets for cancer therapy. *Trends Mol. Med.* **2014**, *20*, 385–393. [CrossRef] [PubMed]
112. Lai, R.C.; Yeo, R.W.; Tan, K.H.; Lim, S.K. Exosomes for drug delivery—A novel application for the mesenchymal stem cell. *Biotechnol. Adv.* **2013**, *31*, 543–551. [CrossRef] [PubMed]
113. Shaw, D.M.; Elger, B.S.; Colledge, F. What is a biobank? Differing definitions among biobank stakeholders. *Clin. Genet.* **2014**, *85*, 223–227. [CrossRef] [PubMed]
114. Artene, S.A.; Ciurea, M.E.; Purcaru, S.O.; Tache, D.E.; Tataranu, L.G.; Lupu, M.; Dricu, A. Biobanking in a constantly developing medical world. *Sci. World J.* **2013**, *2013*, 343275. [CrossRef] [PubMed]
115. Natasha, G.; Gundogan, B.; Tan, A.; Farhatnia, Y.; Wu, W.; Rajadas, J.; Seifalian, A.M. Exosomes as immunotheranostic nanoparticles. *Clin. Ther.* **2014**, *36*, 820–829. [CrossRef] [PubMed]
116. Marcus, M.E.; Leonard, J.N. Fedexosomes: Engineering therapeutic biological nanoparticles that truly deliver. *Pharmaceuticals* **2013**, *6*, 659–680. [CrossRef] [PubMed]
117. Kalra, H.; Adda, C.G.; Liem, M.; Ang, C.S.; Mechler, A.; Simpson, R.J.; Hulett, M.D.; Mathivanan, S. Comparative proteomics evaluation of plasma exosome isolation techniques and assessment of the stability of exosomes in normal human blood plasma. *Proteomics* **2013**, *13*, 3354–3364. [CrossRef] [PubMed]
118. Sokolova, V.; Ludwig, A.K.; Hornung, S.; Rotan, O.; Horn, P.A.; Epple, M.; Giebel, B. Characterisation of exosomes derived from human cells by nanoparticle tracking analysis and scanning electron microscopy. *Colloids Surf. B Biointerfaces* **2011**, *87*, 146–150. [CrossRef] [PubMed]

119. Witwer, K.W.; Buzas, E.I.; Bemis, L.T.; Bora, A.; Lasser, C.; Lotvall, J.; Nolte-'t Hoen, E.N.; Piper, M.G.; Sivaraman, S.; Skog, J.; et al. Standardization of sample collection, isolation and analysis methods in extracellular vesicle research. *J. Extracell. Vesicles* **2013**, *2*, 20360. [CrossRef] [PubMed]
120. Lorincz, A.M.; Timar, C.I.; Marosvari, K.A.; Veres, D.S.; Otrokocsi, L.; Kittel, A.; Ligeti, E. Effect of storage on physical and functional properties of extracellular vesicles derived from neutrophilic granulocytes. *J. Extracell. Vesicles* **2014**, *3*, 25465. [CrossRef] [PubMed]

MDPI
St. Alban-Anlage 66
4052 Basel
Switzerland
Tel. +41 61 683 77 34
Fax +41 61 302 89 18
www.mdpi.com

Journal of Clinical Medicine Editorial Office
E-mail: jcm@mdpi.com
www.mdpi.com/journal/jcm

www.ingramcontent.com/pod-product-compliance
Lightning Source LLC
Chambersburg PA
CBHW042009260926
51850CB00020B/2936